ELECTROENCEPHALOGRAPHY

Edited by **Phakkharawat Sittiprapaporn**

Electroencephalography
http://dx.doi.org/10.5772/65169
Edited by Phakkharawat Sittiprapaporn

Contributors
Foteini Kariotou, Michael Doschoris, Wei-Syuan Chao, Jun-Da Chen, Chin-Feng Lin, Takayuki Kodama, Hideki Nakano, Yasue Mitsukura, Suguru Kanoga, Takashi Murashita, Yuya Sakimoto

Published by InTech
Janeza Trdine 9, 51000 Rijeka, Croatia

Publishing Process Manager Nina Kalinic
Technical Editor SPi Global
Cover InTech Design team

First published December, 2017
Printed in Croatia
Legal deposit, Croatia: National and University Library in Zagreb

Additional hard copies can be obtained from orders@intechopen.com

Electroencephalography, Edited by Phakkharawat Sittiprapaporn
p. cm.
Print ISBN 978-953-51-3637-8
Online ISBN 978-953-51-3638-5

For EU product safety concerns:
IN TECH d.o.o., Prolaz Marije Krucifikse Kozulić 3, 51000 Rijeka, Croatia,
info@intechopen.com or visit our website at intechopen.com.

Meet the editor

Assistant Professor Dr. Phakkharawat Sittiprapaporn is a Thai Cognitive Neuroscientist who has published in many national and international publications. His other achievements include authoring chapters in other books and reviewing articles for reputed national and international indexed journals. Assistant Professor Dr. Phakkharawat Sittiprapaporn was born in Udornthani, Thailand in 1970. He received the B.A. (Hons.) (English) and M.A. (Linguistics) from Srinakharinwirot University and Mahidol University, Bangkok, Thailand in 1993 and 1997, respectively. He received a Ph.D. degree in Neurosciences from Mahidol University, Thailand, in 2002 and became a postdoctoral fellow at the Clinical Cognitive Neuroscience Center (CCNC), Seoul National University, College of Medicine, Seoul, Korea, in 2006. After working as a lecturer in Neurosciences at the Neuro-Behavioural Biology Center, Institute of Science and Technology for Research and Development, Mahidol University, from 2003 to 2005, he was invited to join the Music Therapy Division, College of Music, Mahidol University, Bangkok, in 2008. Since January 2010, he has joined the Faculty of Medicine, Mahasarakham University. He also went for a postdoctoral research fellow at the University of Kansas Medical Center, Kansas, USA. He concurrently held an Assistant Professor at the School of Anti-Aging and Regenerative Medicine, Mae Fah Luang University, Thailand. He also takes in chart of the head of the Brain Sciences and Engineering Research Unit (BSEI), Mae Fah Luang University, Thailand. His current research includes Cognitive Neurosciences, Biomedical Image and Signal Processing of Human Brain Function in Music Perception and Cognition.

Contents

Preface

This edited volume is a collection of reviewed and relevant research chapters, concerning the developments within the electroencephalography field of study. The book includes scholarly contributions by various authors and edited by a group of experts pertinent to medicine and health sciences. Each contribution comes as a separate chapter complete in itself but directly related to the book's topics and objectives.

This book includes chapters dealing with the topics: "Understanding Neural Mechanisms of Action Observation for Improving Human Motor Skill Acquisition," "Hippocampal Theta Activity during Stimulus Discrimination Task," "Intraoperative Electroencephalography during Aortic Arch Surgery," "Mathematical Foundation of Electroencephalography," "Filter Band Multicarrier-Based Transmission Technology for Clinical EEG Signals," and "Review of Artifact Rejection Methods for Electroencephalographic Systems."

The target audience comprises scholars and specialists in the field.

InTechOpen

Chapter 1

Understanding Neural Mechanisms of Action Observation for Improving Human Motor Skill Acquisition

Hideki Nakano and Takayuki Kodama

Additional information is available at the end of the chapter

http://dx.doi.org/10.5772/intechopen.69266

Abstract

Action observation is a useful approach for improving human motor skill acquisition. This process involves the mirror neuron system that consists of the ventral premotor area, inferior parietal lobule, and superior temporal sulcus. The interaction between these areas produces the effect of action observation. This chapter presents neurophysiological and brain imaging studies of action observation, and their application to human motor learning. For action observation, the mirror system appears to map the intention in the ventral premotor area and the goal in the inferior parietal lobule. These features of action representation may be useful for refining conditions of practice, based on the mirror system, for acquiring new motor skills.

Keywords: action observation, electroencephalography, mirror neuron system, motor learning

1. Introduction

Previous neurophysiological and brain imaging studies have revealed that neural activity associated with observation of another person's movement was elicited in the motor-related cortical areas [1–3]. The motor-related cortical areas that get active during such motion perception, constitute the mirror neuron system. Characteristically, this system is activated not only when a person performs a goal-oriented movement by himself/herself, but also when the person observes the same movement performed by others (**Figure 1**) [4]. Action observation automatically creates a similar simulation of movement in the brain of the observer [5, 6]. In other words, action observation induces functional reorganization of the brain, and facilitates motor learning via the mirror neuron system (**Figure 2**) [7, 8]. Thus, observation of another person's actions and behavior alters the neuronal activity of the observer. This chapter discusses neurophysiological and brain imaging studies of action observation, and their application to human motor learning.

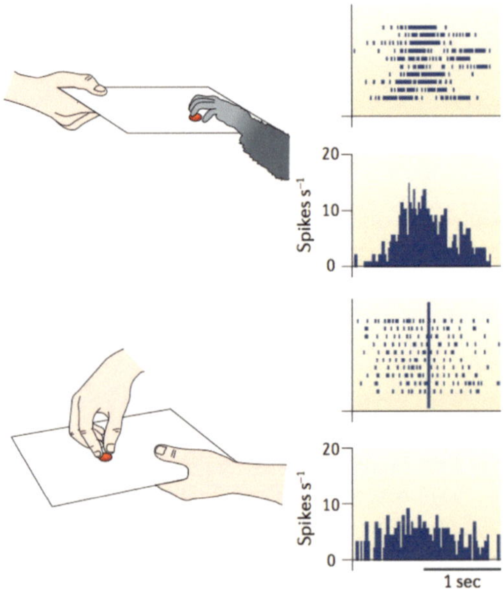

Figure 1. Mirror neurons in monkeys [4]. Top indicates the neural activity in area F5 when the monkey grasps food. Bottom indicates the neural activity in are F5 when the monkey observes the human grasping food.

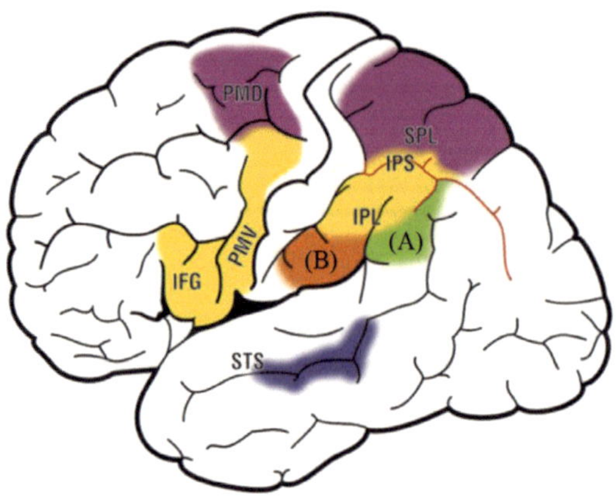

Figure 2. Mirror neuron system in humans [7]. Purple areas (PMD and SPL) are involved in reaching movements. Yellow areas (IFG, PMV, IPL and IPS) are involved in transitive distal movements. Blue areas (STS) are involved in observation of upper-limb movements. Green areas (A) are involved in intransitive movements. Orange areas (B) are involved in tool use. PMD indicates dorsal premotor cortex; SPL, superior parietal lobule; IFG, inferior frontal gyrus; PMV, ventral premotor cortex; IPL, inferior parietal lobule; IPS, intraparietal sulcus; STS, superior temporal sulcus.

http://dx.doi.org/10.5772/intechopen.69266

2. Neural mechanism of action observation

The mirror neuron system is concerned with the neural mechanism of action observation. The mirror neuron system is a neural system that consists of the ventral premotor area, inferior parietal lobule, and superior temporal sulcus, and the interaction between these areas produces the effect of action observation (**Figure 3**) [9]. The mirror neuron system converts sensory information obtained through observation to a specific motion pattern, which is an objective [10]. Thus, decryption of the action and behavior of other individuals are facilitated. Additionally, the encoding of the meaning and objective of an action facilitates understanding of the purpose of the action performed by others, internally, without requiring higher cognitive processes, such as reasoning. The understanding of other's intentions [11], empathy [12], and theory of mind [13] are known functional characteristics of the mirror neuron system, and dysfunction of the mirror neuron system is associated with autism [14].

Fadiga et al. investigated the mirror neuron system using transcranial magnetic stimulation (TMS) [5]. They measured the motor evoked potential (MEP), which is the index of excitatory change of the corticospinal tract, from finger muscles, by stimulating the motor area by means of TMS, while the subject was observing the action. The results showed that the MEP amplitude of the finger muscle involved in the observed action, increased significantly, and these phenomena were not observed in the MEP amplitude of another finger muscle that was not concerned with the observed action. Thus, the peripheral motor system also prepares for performing the observed action, and the temporal consistency between the muscle group concerned with the action that is an objective and the muscle activation patterns indicate that the mirror neuron system couples action execution and observation [15]. The same motor representations are activated during both action execution and observation; this indicates the possibility that the reorganization of the motor system network that is induced by action observation and that which is induced by actual physical practice involve the same mechanisms [16].

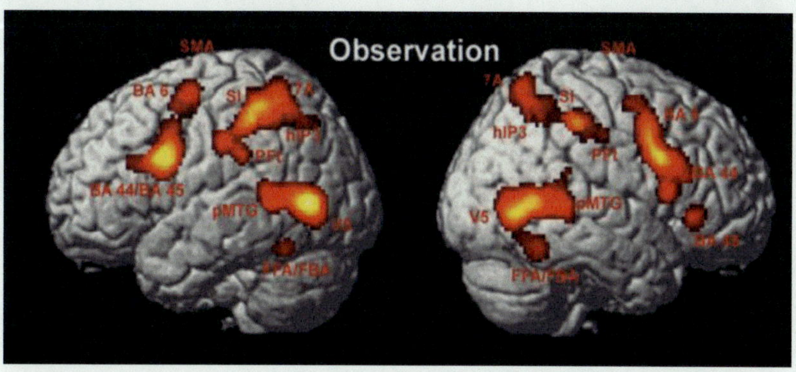

Figure 3. ALE meta-analysis of action observation in the human brain [9]. Significant meta-analysis results for action observation, summarized over all effectors. All results are displayed on the left and right lateral surface view of the MNI single subject template.

Many previous studies have investigated the excitatory changes in the corticospinal tract during a single action observation, and some of these have examined the excitatory change in the corticospinal tract during repetitive action observation. Stefan et al. revealed that repetitive action observation changes the motor representation in the cerebral cortex as it does with physical practice [17]. Moreover, they disclosed that these changes of the motor representation in the cerebral cortex, were induced by physical practice along with brief action observation. Furthermore, it has been reported that repetitive action observation increased the excitability of the corticospinal tract, and that there was a significant positive correlation between the increased excitability of the corticospinal tract and the change in motion patterns [16].

Watanabe et al. examined the effect of observation viewpoint on brain activity and performance [18]. There are two observation perspectives. One involves observing another person's action from the same side as that of the subject's perspective, termed the "first-person perspective" and the other involves observing the other person's action on the opposite side as that of the subject's perspective, termed the "third-person perspective" (**Figure 4**). Their study investigated the difference between reaction time and brain activity from first-person and third-person perspectives, while the subject observed the action and imitated the action of another, using functional magnetic resonance imaging (fMRI). They showed that the motor-related areas involve the mirror neuron system, including the ventral premotor area, supramarginal gyrus, and supplementary motor area, in the first-person perspective significantly more than in the third-person perspective. Thus, action observation from the first-person perspective activates the mirror neuron system advantageously, and facilitates the intracerebral simulation of the action that is the objective, and enhances motor learning. This information should be applied to motor learning, considering the various conditions of action observation.

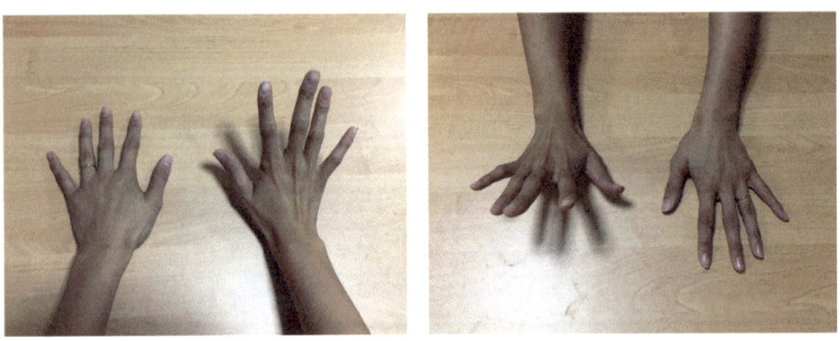

Figure 4. First-person perspective (left) and third-person perspective (right).

3. Mirror neuron system and EEG studies

Many studies have investigated the mirror neuron system using electroencephalography (EEG) [19]. A specific EEG rhythm, called the mu rhythm (8–13 Hz), is observed in the human sensorimotor cortex [20]. The characteristics of the mu rhythm are blocked during actual movement

[20] as well as observation of action [21] and motor imagery [22]. Therefore, numerous EEG studies have used the mu rhythm as an electrophysiological marker of the mirror neuron system in humans, after Altschuler et al. [23] first investigated this possibility. Those studies found that the mu rhythm represents the activity of the mirror mechanism in humans. An fMRI study also showed a significant correlation between mu rhythm desynchronization (**Figure 5**) [24] and BOLD activity in typical mirror neuron system regions.

Many studies have examined the reactivity of the mu rhythm during action observation. Avanzini et al. [25] investigated the dynamics of sensorimotor cortical oscillations during

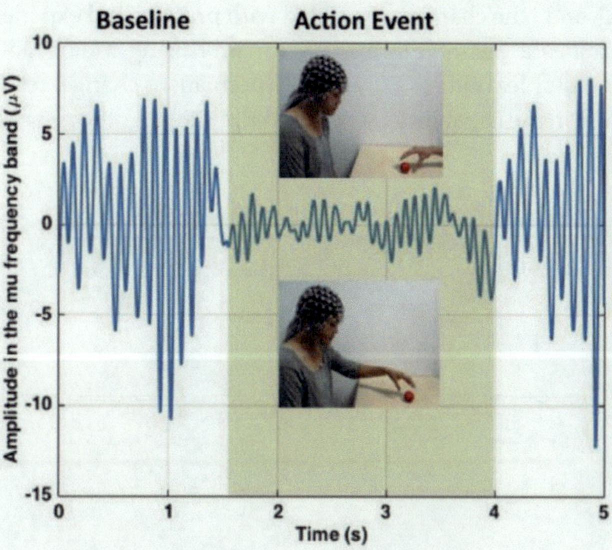

Figure 5. Mu rhythm desynchronization [24]. Simulation of mu rhythm desynchronization in the 8–13 Hz frequency band. There is a decrease in amplitude in the electroencephalogram from baseline during action observation or execution.

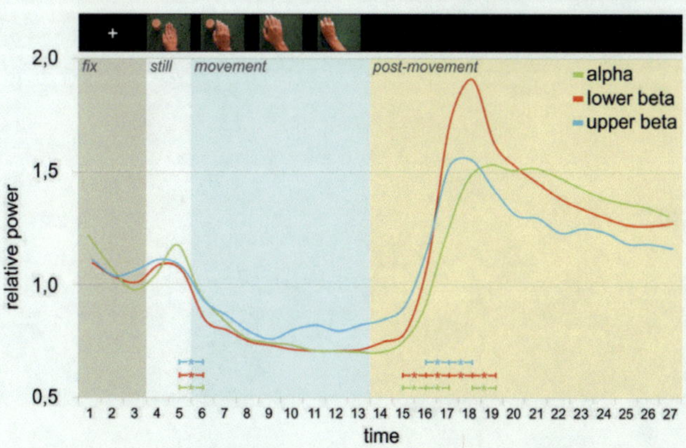

Figure 6. EEG rhythms during action observation [25]. The graph shows the EEG power time-course for each frequency band: alpha band (8–13 Hz) in green, lower beta (13–18 Hz) in red, and upper beta (18–25 Hz) in cyan.

the observation of hand movements using EEG (**Figure 6**). A desynchronization of alpha and beta rhythms was observed in the central and parietal regions. Notably, there was a large post-stimulus power rebound present in all bands. Furthermore, the velocity profile of the observed movement and beta band modulation correlated, indicating a direct matching of the stimulus parameter to motor activity.

4. Action observation and motor learning

Schmidt defined that motor learning is a "process of acquiring the capability for producing skilled actions" [26], and "the changes associated with practice and experiences, in an internal process that determines a person's capability for producing motor skill" [27]. Moreover, Guthrie stated that motor learning is a "relatively permanent change, resulting from practice or a novel experience, in the capability for responding" [28]. In other words, motor learning is

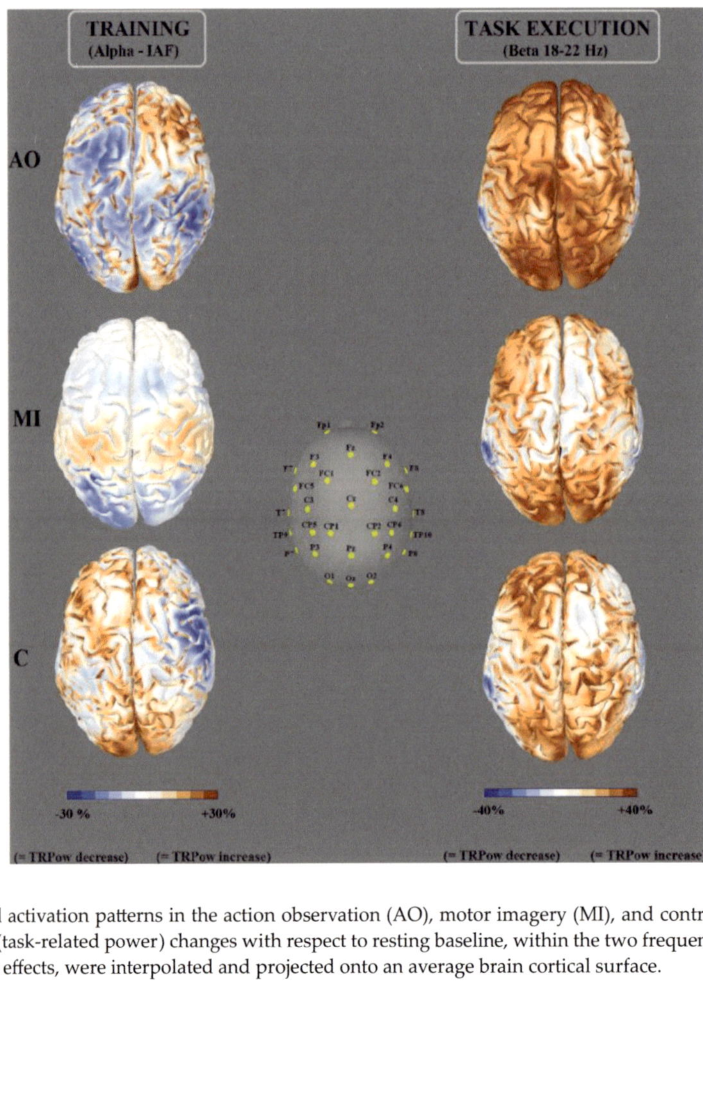

Figure 7. Cortical activation patterns in the action observation (AO), motor imagery (MI), and control (C) groups [34]. Average TRPow (task-related power) changes with respect to resting baseline, within the two frequency bands showing significant group effects, were interpolated and projected onto an average brain cortical surface.

http://dx.doi.org/10.5772/intechopen.69266

the capability acquired from practice and experience, and the change is relatively permanent. How can then action observation contribute to motor learning?

Motor imagery is similar to action observation. It is the mental simulation of movement without physical movement of body parts [29]. Action observation and motor imagery have been shown to share the same neural basis as that used for the execution of the actual physical movement [30]. However, the process of action observation and motor imagery are different. Action observation is a bottom-up process (process from perception), while motor imagery is a top-down process (process from memory). Nevertheless, these processes are not clearly divided, and a feed-forward model was constructed by complementing these processes [31]. In addition, action observation has an effect of promoting motor imagery [32].

In the early stages of new complex motor learning, action observation is superior to motor imagery as a strategy for motor learning, as revealed by behavioral [33] and EEG data (**Figure 7**) [34]. Motor imagery is influenced by the environment and personal imaging ability and requires mental effort. In contrast, action observation is easier to apply than motor imagery, despite targeting activation of the same neural network as motor imagery [35]. We have also reported that the left sensorimotor and parietal areas of the high-motor learning group showed a greater decrease in the alpha-2 and beta-2 rhythms than those of the low-motor learning group during observation and execution. These results suggested that the decreases in the alpha-2 and beta-2 rhythms in these areas during observation and execution

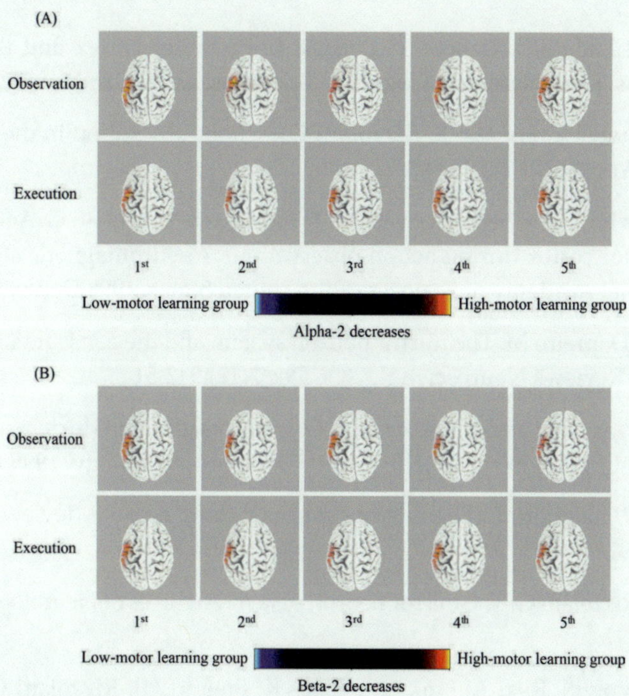

Figure 8. Changes in EEG activity durings observation and execution of a motor learning task [36]. Alpha-2 (A) and beta-2 (B) rhythms in the left sensorimotor and parietal areas during action execution from the 1st trial to the 5th trial were significantly decreased in the high-motor learning group compared with in the low-motor learning group.

are associated with motor skill improvement (**Figure 8**) [36]. Accordingly, action observation may be an effective tool as an intervention method during the early stage of motor learning.

Acknowledgements

This work was supported by JSPS KAKENHI Grant Number JP26750208 and ASTEM RI / KYOTO.

Author details

Hideki Nakano* and Takayuki Kodama

*Address all correspondence to: nakano-h@tachibana-u.ac.jp

Department of Physical Therapy, Faculty of Health Sciences, Kyoto Tachibana University, Kyoto, Japan

References

[1] Rizzolatti G, Fadiga L, Gallese V, Fogassi L. Premotor cortex and the recognition of motor actions. Brain Research. Cognitive Brain Research. 1996 Mar;**3**(2):131-41

[2] Gallese V, Fadiga L, Fogassi L, Rizzolatti G. Action recognition in the premotor cortex. Brain. 1996 Apr;**119**(Pt 2):593-609

[3] Hari R, Forss N, Avikainen S, Kirveskari E, Salenius S, Rizzolatti G. Activation of human primary motor cortex during action observation: A neuromagnetic study. Proceedings of the National Academy of Sciences of the United States. 1998 Dec 8;**95**(25):15061-15065

[4] Iacoboni M, Dapretto M. The mirror neuron system and the consequences of its dysfunction. Nature Reviews Neuroscience. 2006 Dec;**7**(12):942-51

[5] Fadiga L, Fogassi L, Pavesi G, Rizzolatti G. Motor facilitation during action observation: A magnetic stimulation study. The Journal of Neurophysiology. 1995 Jun;**73**(6):2608-11

[6] Rizzolatti G, Craighero L. The mirror-neuron system. Annual Review of Neuroscience. 2004;**27**:169-92

[7] Cattaneo L, Rizzolatti G. The mirror neuron system. Archives of Neurology. 2009 May;**66**(5): 557-60

[8] Buccino G, Vogt S, Ritzl A, Fink GR, Zilles K, Freund HJ, Rizzolatti G. Neural circuits underlying imitation learning of hand actions: An event-related fMRI study. Neuron. 2004 Apr 22;**42**(2):323-34

[9] Caspers S, Zilles K, Laird AR, Eickhoff SB. ALE meta-analysis of action observation and imitation in the human brain. Neuroimage. 2010 Apr 15;**50**(3):1148-67

[10] Rizzolatti G, Fogassi L, Gallese V. Neurophysiological mechanisms underlying the understanding and imitation of action. Nature Reviews Neuroscience. 2001 Sep;**2**(9):661-70

[11] Fogassi L, Ferrari PF, Gesierich B, Rozzi S, Chersi F, Rizzolatti G. Parietal lobe: From action organization to intention understanding. Science. 2005 Apr 29;**308**(5722):662-7

[12] Singer T, Seymour B, O'Doherty J, Kaube H, Dolan RJ, Frith CD. Empathy for pain involves the affective but not sensory components of pain. Science. 2004 Feb 20;**303**(5661):1157-62

[13] Gallese V, Goldman A. Mirror neurons and the simulation theory of mind-reading. Trends in Cognitive Sciences. 1998 Dec 1;**2**(12):493-501

[14] Oberman LM, Hubbard EM, McCleery JP, Altschuler EL, Ramachandran VS, Pineda JA. EEG evidence for mirror neuron dysfunction in autism spectrum disorders. Brain Research. Cognitive Brain Research. 2005 Jul;**24**(2):190-8

[15] Small SL, Buccino G, Solodkin A. The mirror neuron system and treatment of stroke. Developmental Psychobiology. 2012 Apr;**54**(3):293-310

[16] Ray M, Dewey D, Kooistra L, Welsh TN. The relationship between the motor system activation during action observation and adaptation in the motor system following repeated action observation. Human Movement Science. 2013 Jun;**32**(3):400-11

[17] Stefan K, Cohen LG, Duque J, Mazzocchio R, Celnik P, Sawaki L, Ungerleider L, Classen J. Formation of a motor memory by action observation. The Journal of Neuroscience. 2005 Oct 12;**25**(41):9339-46

[18] Watanabe R, Higuchi T, Kikuchi Y. Imitation behavior is sensitive to visual perspective of the model: An fMRI study. Experimental Brain Research. 2013 Jul;**228**(2):161-71

[19] Rizzolatti G, Cattaneo L, Fabbri-Destro M, Rozzi S. Cortical mechanisms underlying the organization of goal-directed actions and mirror neuron-based action understanding. Physiological Reviews. 2014 Apr;**94**(2):655-706

[20] Gastaut H, Terzian H, Gastaut Y. Etude d'une activite electroencephalographique meconnue: le rythme rolandique en arceau. Mars Med. 1952;**89**(6):296-310

[21] Gastaut HJ, Bert J. EEG changes during cinematographic presentation; moving picture activation of the EEG. Electroencephalography and Clinical Neurophysiology. 1954 Aug;**6**(3):433-44

[22] Pfurtscheller G, Neuper C. Motor imagery activates primary sensorimotor area in humans. Neuroscience Letters. 1997 Dec 19;**239**(2-3):65-8

[23] Altschuler EL, Vankov A, Wang V, Ramachandran VS, Pineda JA. Person see, person do: Human cortical electrophysiological correlates of monkey see monkey do cells. Poster Session Presented at the 27th Annual Meeting of the Society for Neuroscience; New Orleans: LA. 1997

[24] Fox NA, Bakermans-Kranenburg MJ, Yoo KH, Bowman LC, Cannon EN, Vanderwert RE, Ferrari PF, van IJzendoorn MH. Assessing human mirror activity with EEG mu rhythm: A meta-analysis. The Psychological Bulletin. 2016 Mar;**142**(3):291-313

[25] Avanzini P, Fabbri-Destro M, Dalla Volta R, Daprati E, Rizzolatti G, Cantalupo G. The dynamics of sensorimotor cortical oscillations during the observation of hand movements: an EEG study. PLoS One. 2012;**7**(5):e37534

[26] Schmidt RA. Motor Control and Learning: A Behavioral Emphasis. 2nd ed. Champaign: Human Kinetics Publisher; 1988. p. 346

[27] Schmidt RA. Motor Learning & Performance: From Principles to Practice. Champaign: Human Kinetics Publisher; 1991. p. 51

[28] Guthrie ER. The Psychology of Learning. New York: Harper & Row; 1952

[29] Decety J, Perani D, Jeannerod M, Bettinardi V, Tadary B, Woods R, Mazziotta JC, Fazio F. Mapping motor representations with positron emission tomography. Nature. 1994 Oct 13;**371**(6498):600-2

[30] Grèzes J, Decety J. Functional anatomy of execution, mental simulation, observation, and verb generation of actions: a meta-analysis. Human Brain Mapping. 2001 Jan;**12**(1):1-19

[31] Holmes P, Calmels C. A neuroscientific review of imagery and observation use in sport. Journal of Motor Behavior. 2008 Sep;**40**(5):433-45

[32] Conson M, Sarà M, Pistoia F, Trojano L. Action observation improves motor imagery: Specific interactions between simulative processes. Experimental Brain Research. 2009 Oct;**199**(1):71-81

[33] Gatti R, Tettamanti A, Gough PM, Riboldi E, Marinoni L, Buccino G. Action observation versus motor imagery in learning a complex motor task: A short review of literature and a kinematics study. Neuroscience Letters. 2013 Apr 12;**540**:37-42

[34] Gonzalez-Rosa JJ, Natali F, Tettamanti A, Cursi M, Velikova S, Comi G, Gatti R, Leocani L. Action observation and motor imagery in performance of complex movements: Evidence from EEG and kinematics analysis. Behavioural Brain Research. 2015 Mar 15;**281**:290-300

[35] Buccino G. Action observation treatment: A novel tool in neuro rehabilitation. Philosophical transactions of the Royal Society of London. Series B, Biological Sciences. 2014 Apr 28;**369**(1644):20130185

[36] Nakano H, Osumi M, Ueta K, Kodama T, Morioka S. Changes in electroencephalographic activity during observation, preparation, and execution of a motor learning task. International Journal of Neuroscience. 2013 Dec;**123**(12):866-75

Chapter 2

Hippocampal Theta Activity During Stimulus Discrimination Task

Yuya Sakimoto and Dai Mitsushima

Additional information is available at the end of the chapter

http://dx.doi.org/10.5772/intechopen.68461

Abstract

The configural association theory and conflict resolution model both propose that hippocampal function plays role in the solving a negative patterning task but not simple discrimination task. Some hippocampal lesion study showed that inactivity of rats' hippocampal CA1 area induced impairment of performance of a negative patterning task. Other previous studies, however, showed that the lesion did not affect the performance of the task. Thus, it did not reveal whether hippocampal function was important for solving the negative patterning task. Our recent research using an electrophysiological approach showed that the hippocampal theta power decreased with a compound stimulus of a negative patterning task, and that the hippocampal theta power was decreased by a compound stimulus of a feature negative task. These results indicate that a decrease in hippocampal theta activity is elicited by behavioral inhibition for conflict stimuli with overlapping elements. This finding strongly supports the conflict resolution model and suggests a hippocampal role in learning behavioral inhibition for conflict stimuli during nonspatial stimulus discrimination tasks.

Keywords: hippocampal theta power, negative patterning task, feature negative task, configural association theory, conflict resolution model

1. Hippocampal memory functions

After report of a patient H.M. [1], the various areas of research, including psychology, neuroscience, cognitive science, and behavioral science, have researched the hippocampal function for learning and memory. The patient H.M. suffered epileptic seizures as a child. He receives a resection of a portion of temporal lobe including hippocampus for treating the seizures when he was 27 years old. Although the treatment reduced the symptom of seizure without lack of some his cognitive and memory functions, such as intelligence quotient (IQ), conversational ability, perceptual capability, working memory, semantic memory formed before surgery, and procedure memory, the patient H.M. had minor symptoms of retrograde amnesia and severs anterograde amnesia [1]. Thus, Scoville and Milner [1] suggested that the hippocampus is necessary for the encoding of episode memory but not retrieval and storage of the memory. After then, for understanding the hippocampal function in detail, animal researchers have examined what kind of learning task is necessary for being solved by hippocampal functions. In the electrophysiology study, O'Keefe and Nadel [2] showed that the rodents' hippocampal CA1 neuron was activated by the memory of placement. They named "place cell" as hippocampal CA1 neuron which associated information of placement and suggested the cognitive theory that hippocampus was important for solving spatial learning by using eight arms radical maze and Morris water maze. In the hippocampal lesion study, Bouffard and Jarrard [3] compared hippocampal lesion rat and control rat without hippocampal lesion on eight arms radical maze. For solving this task, rats need to learn the arms that were choose once time by using peripheral environmental cue outside of the maze. The performance of the rats with hippocampal lesion was less than that control rats. Also, it has examined on the Morris water maze. For solving the task, the rats need to understand own position and goal position from some environmental stimuli outside the maze and reach the goal position by cueing these stimuli. The rats with hippocampal lesion increase a latency that reached an invisible goal platform as compared with control rats on the maze [4]. In addition, several research studies showed the universal function that hippocampus plays role in a spatial learning over other species, such as fishes [5], birds [6, 7], and primate [8, 9], suggesting that the cognitive map theory is one of the popular hippocampal function theories having adaptive possibility for various species. On the other hand, some researchers have reported that hippocampal function was important for solving a certain type of nonspatial stimulus discrimination task.

2. Negative patterning task

Configural association theory suggests that the hippocampus plays role in learning the relationship between multiple sensory stimuli [10]. According to the theory, animals have two systems, elemental and configural association systems, for processing sensory information, and they adapt successfully to various situations in the external world by using them. The elemental association system forms representation of single stimulus, such as the single stimulus associated with reinforcement or punishment. However, in the external world, a compound stimulus combining multiple stimuli may sometimes have a significant meaning. The configural association system forms one of the configural representations by associating between multiple stimuli when some of them are presented simultaneously or serially. Sutherland and Rudy [10] proposed that hippocampal function was necessary for the formation of configural representations for compound stimuli. After then, the theory was revised by some researchers [11–13] and latest theory that the hippocampus is important for configural presentation for compound stimulus in exclusive-or (XOR) tasks such as negative patterning task and positive patterning task. In the negative patterning task, rats are reinforced for operant responses when either one of two different sensory single stimuli, such as tone (T) or light (L), is presented (T+ or L+). In contrast, rats are not reinforced when both stimuli are presented (TL−). In this task, compound stimulus had overlapping element with single stimuli. For solving this task, rats need to, thus, form the configural representation for compound stimulus and discriminate between compound stimulus and single stimuli.

Gray and McNaughton [14] proposed the conflict resolution model for the hippocampal function and behavioral inhibition. The model suggests that the hippocampal function plays role in the resolution of conflict between incompatible goals or response tendencies. According to this model, the hippocampal function modulates the weight of negative information and, specifically, increases it, thereby inducing behavioral inhibition [14, 15]. Interestingly, this theory may also explain why the hippocampus is important for solving the compound stimulus in the negative patterning task. In the negative patterning task, either one of stimuli A and B is presented alone when they signal a "go" response, but the stimuli are presented simultaneously when they signal a "no-go" response. Thus, the compound stimulus had incompatible goals or response tendencies. Animals need to increase the weight of negative information and inhibit operant response for compound stimulus. Both the conflict resolution model and the configural association theory suggest a role of the hippocampus in solving the negative patterning task.

3. Hippocampal lesion and negative patterning task

Several previous studies reported that hippocampal lesions impair the negative patterning task performance [10, 16–18]. However, some studies have reported no effects of hippocampal lesions on the negative patterning task [19]. Davidson et al. [19] reported abnormal behavioral inhibition of the operant response after hippocampal lesions, suggesting a lack of learning ability for compound stimuli. Moreover, hippocampal lesions result in an abnormal operant response, such as response persistence [20]. Therefore, in order to build a more solid foundation for the configural association theory and conflict resolution model, the relationship between hippocampal activity and the negative patterning task needs to be investigated by means other than hippocampal lesions.

4. Hippocampal theta activity during negative patterning

It has been known that electroencephalography (EEG) was useful for neural activity of hippocampus without extensive hippocampal lesions in rodents. When we implanted a recording polar into the rats' hippocampal CA1 area, we can observe rhythmic EEG patterns. The EEG activity was grouped: theta waves (6–12 Hz), beta waves (12–30 Hz), gamma waves (30–100 Hz), and ripple waves (100–200 Hz). Specifically, it was known that hippocampal theta wave is related to psychological state and behavior. Several studies have reported that hippocampal theta waves strongly are related to locomotor behavior such as running, jumping, ricking, and operant response [21–24]. In addition, it has reported that hippocampus theta activity is related to learning and memory [25–43]. Masuoka et al. [29] showed that rats' hippocampal theta activity increased during elevated radical eight mazes. Also, Olvera-Cortés et al. [30, 31] revealed that the hippocampal theta waves change during performance of spatial learning task. Thus, in addition, theta waves are thought to occur by the synchronization of neurons in the whole hippocampal formation [44], which would reflect hippocampal neural activity [45–48].

Recently, we have examined hippocampal theta activity in rats during the acquisition stages (early, middle, and late) of the negative patterning task (T+, L+, TL−) [38]. We observed a transient decrease in hippocampal theta power immediately after the presentation of a compound stimulus during the late stage of learning in the negative patterning task (**Figure 1**). In addition, the magnitude of the decrease in theta power strongly correlated with improved performance in the negative patterning task (**Figure 2**). Grastyán et al. [49] examined the relationship between hippocampal theta activity and the acquisition of an orientative conditioned response (CR) for a tone stimulus presentation in cats. Although the hippocampal theta activity increased with an association between stimulus and orientative CR, the hippocampal theta wave decreased after the formation of this association. Thus, the transient decrease in hippocampal theta activity during the late stage of learning in the negative patterning task observed in the current study may be related to mastery of the negative patterning task. However, our previous reports showed a greater decrease in hippocampal theta activity in the late learning stage of a negative patterning task compared to the simple discrimination task [32, 38]. Therefore, we suggest that the decrease in hippocampal theta power is induced by hippocampus-mediated information processing for compound stimuli in the negative patterning task. This is in agreement with the concepts of the configural association theory and conflict resolution model.

Further studies revealed characteristics of compound stimuli inducing a decrease in theta power by comparing simultaneous feature-negative (T+, TL−) and compound stimulus discrimination tasks (T_1+, T_2L−) [32]. In feature negative tasks, the compound stimulus had overlapping elements with single stimuli because these stimuli were composed of tone stimuli with the same frequency component. However, the compound stimulus in the compound stimulus discrimination task did not have overlapping elements with single stimuli because they were composed of tone stimuli with different frequency components (T_1: 2000 Hz, T_2: 4000 Hz). These studies reported a transient decrease in hippocampal theta activity following the presentation of a compound stimulus during the simultaneous feature-negative task compared to the simple discrimination task but not during the compound stimulus discrimination task. The compound stimulus of the simultaneous feature-negative task had an overlapping element shared with the single stimulus. This may justify the transient decrease in hippocampal theta activity during response inhibition for the compound stimulus of negative patterning and simultaneous feature-negative tasks. Therefore, we proposed that the decrease in hippocampal theta power is related to behavioral inhibition for conflict stimulus discrimination in which the single stimuli have overlapping elements.

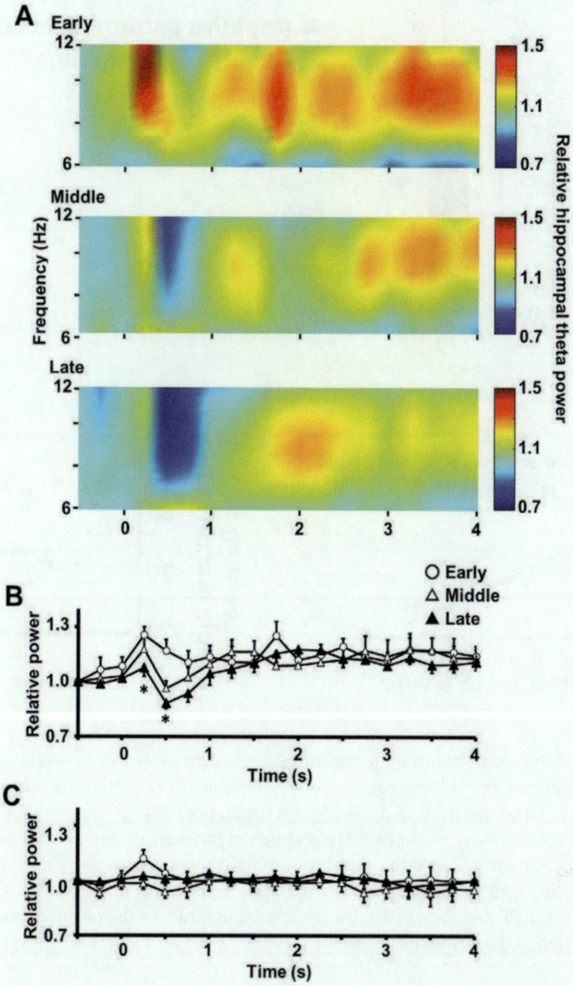

Figure 1. The change in theta power during a presentation of compound stimuli of the negative patterning task by using wavelet analysis (A). Upper side shows the change in hippocampal theta activity along a time course during compound stimulus on the early stage, Middle side shows theta activity on the middle stage, and Lower side shows theta activity on late stage of negative patterning task. The x-axis is time (ms), and the y-axis is frequency (Hz). In each panel, the period is from 500 ms before stimulus onset to 4000 ms after stimulus onset. The mean hippocampal theta power during 500 ms before stimulus onset was counted as the −500-ms period (no stimuli were present and no rats pressed the lever during this period), and the relative theta power calculated for each period (per 250 ms) was normalized to that during the −500-ms period (relative theta activity of each period = theta power of each period/theta power at the −500-ms period). Panel B contains a comparison of the mean (± S.E.M.) relative hippocampal theta activity at 6–12 Hz among each learning stage (early, middle, and late) throughout the time course of the experiment during compound stimuli of the negative patterning task. Two-way within-subjects ANOVA suggests that there is a significant interaction of learning stages (early, middle, and late) × epochs (−500 to 4000 ms, with each 250 ms; $F(36,180) = 2.37$, $p < 0.05$) and a significant effect of epochs ($F(18,90) = 4.80$, $p < 0.05$), but no significant effect of stages ($F(2,10) = 0.97$, n.s.) on relative hippocampal theta power during compound stimulus of the negative patterning task. *Post-hoc* tests showed that there was a significant simple main effect in the 250- and 500-ms epochs during compound stimulus. Multiple comparisons revealed that hippocampal theta power decreased in the 250-ms epochs during nonRFTs in the late stage compared with the early stage ($p < 0.05$) and in the 500-ms epochs during nonRFTs in the middle and late stages compared with the early stage (* $p < 0.05$). Panel C contains a comparison of the mean (± S.E.M.) relative hippocampal theta activity at 6–12 Hz among each learning stage (early, middle, and late) throughout the time course of the experiment during nonreinforced stimulus of the simple discrimination task. This figure was referred to Sakimoto et al. [38].

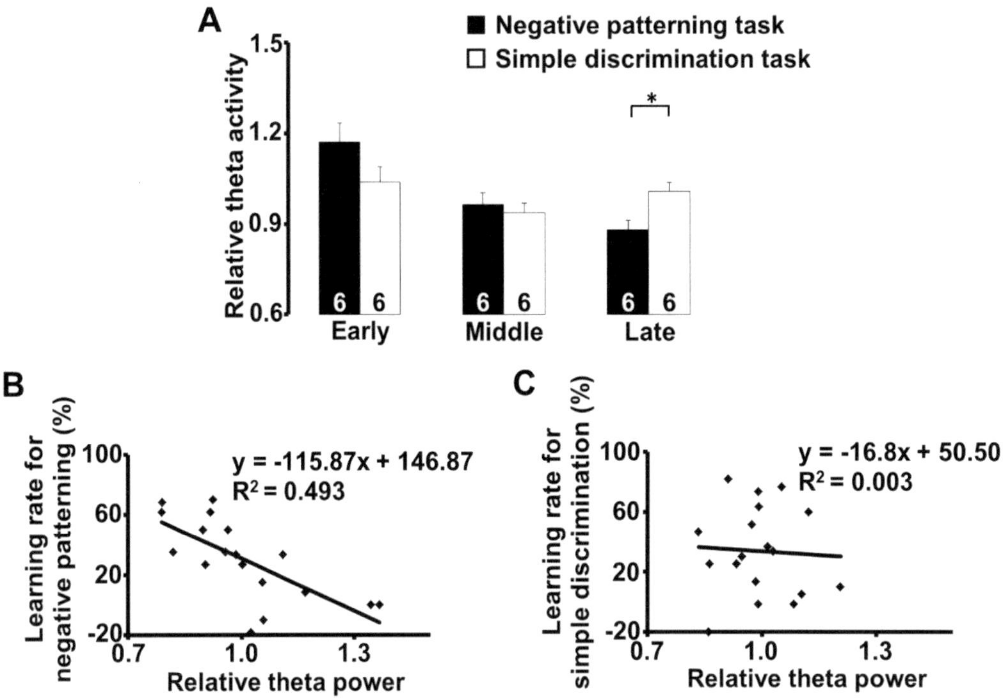

Figure 2. A comparison of the mean relative hippocampal theta activity between tasks. Panel A shows the relative hippocampal theta power during the 500-ms epochs between the negative patterning and simple discrimination task groups. A group (negative patterning task and simple discrimination task groups) × stage (early, middle, and late) ANOVA for hippocampal theta activity during a 500-ms epoch in the nonRFT showed a significant interaction ($F(2,20) = 6.12$, $p < 0.05$). Multiple comparisons revealed that hippocampal theta power decreased during the late stage in the negative patterning task compared to the simple discrimination task group ($p < 0.05$; *: $p < 0.05$). Hippocampal theta power during the 500 ms nonRFT correlated with the discrimination rate in the negative patterning task ($r = -0.70$, $p < 0.05$; panel B) but not the simple discrimination task ($r = -0.06$, p = n.s; panel C). This figure was referred to Sakimoto et al. [38].

5. Why does hippocampal theta amplitude decline?

Hippocampal theta power is affected by the activity of cholinergic and γ-aminobutyric acid (GABA) neurons of the medial septal/diagonal band area [44]. Monmaur and Breton [50] demonstrated that theta activity increases when the cholinergic agonist, carbachol, is injected into the intra septum in freely moving rats. In addition, Sun et al. [51] reported that hippocampal theta activity is abolished by the GABA antagonist bicuculline. Thus, we propose that the transient decrease in hippocampal theta activity during compound stimulus learning in the negative patterning task is induced by the activity of septal cholinergic or GABAergic neurons, or their interaction. In future studies, the relationship between the negative patterning task and septal cholinergic and/or GABAergic activity should be examined. Because septo-hippocampal GABAergic input to CA1 is essential for the generation of theta waves [52], the transient decrease

http://dx.doi.org/10.5772/intechopen.68461

in presynaptic GABA release may cause a transient decrease in the hippocampal theta power immediately after stimulus presentation. We recently analyzed synaptic plasticity using the slice patch-clamp technique and measured a rapid decrease in presynaptic GABA release at hippocampal CA1 synapses immediately after the nonspatial contextual learning task, inhibitory avoidance (IA) [53]. Compared to untrained controls, the paired pulse ratio (PPR) of evoked inhibitory postsynaptic current (IPSC) increased immediately after IA training (at 0 min), suggesting an acute decrease in the probability of presynaptic GABA release. As the PPR returned to baseline 5 min after the training, the decrease in presynaptic GABA release seems to be transient. Moreover, we observed a sustained increase in the miniature excitatory postsynaptic current (mEPSC) and miniature inhibitory postsynaptic current (mIPSC) amplitudes 5–30 min after the IA task, suggesting long-term postsynaptic strengthening of α-amino-3-hydroxy-5-methyl-4-isoxazole propionic acid (AMPA) and $GABA_A$ receptor-mediated synapses. In addition, the long-term increase in mIPSC frequency is probably due to an increase in the number of $GABA_A$ receptor-mediated inhibitory synapses after the training [53].

6. Conclusion

We discussed hippocampal function in a nonspatial stimulus discrimination task with a focus on the configural association theory and conflict resolution model. These functions were strongly supported by the observations that hippocampal theta power decreased during the presentation of a compound stimulus in a negative patterning task [32, 38]. A transient decrease in hippocampal theta activity was also observed during the presentation of a compound stimulus in the simultaneous feature-negative task but not in the compound stimulus discrimination task [32]. These results suggest that the decrease in hippocampal theta activity was elicited by behavioral inhibition of a conflict stimulus with overlapping elements. Therefore, we conclude that the hippocampus may play a role in this cognitive process. This conclusion strongly supports the conflict resolution model, in which the hippocampus plays a role in negative information processing for conflict stimuli in the nonspatial discrimination task. Moreover, data suggest a link between a decreased in theta power and decreased septal cholinergic activity and increased septal GABAergic activity. Finally, we conclude that the hippocampal neural activity derived from septal cholinergic and GABAergic activities plays a central role in behavioral inhibition for conflict stimuli with overlapping elements.

Author details

Yuya Sakimoto* and Dai Mitsushima

*Address all correspondence to: y-saki@yamaguchi-u.ac.jp

Department of Physiology, Graduate School of Medicine, Yamaguchi University, Minami-Kogushi, Ube, Japan

References

[1] Scoville WB, Milner B. Loss of recent memory after bilateral hippocampal lesions. Journal of Neurology Neurosurgery Psychiatry. 1957;**20**:11-21

[2] O'Keefe J, Nadel L. The Hippocampus as a Cognitive Map. Oxford, UK: Oxford University Press; 1978

[3] Bouffard JP, Jarrard LE. Acquisition of a complex place task in rats with selective ibotenate lesions of hippocampal formation: Combined lesions of subiculum and entorhinal cortex versus hippocampus. Behavioral Neuroscience. 1988;**102**:828-834

[4] Gallagher M, Holland PC. Preserved configural learning and spatial learning impairment in rats with hippocampal damage. Hippocampus. 1992;**2**:81-88

[5] Saito K, Watanabe S. Experimental analysis of spatial learning in goldfish. Psychological Record. 2005;**55**:647-662

[6] Watanabe S, Bischof H-J. Effects of hippocampal lesions on acquisition and retention of spatial learning in zebra finches. Behavioral Brain Research. 2004;**155**:147-152

[7] Pearce JM, George DN, Hallselgrove M, Erichsen JT, Good MA. The influence of hippocampal lesions on the discrimination of structure and on spatial memory in pigeons (*Columba livia*). Behavioral Neuroscience. 2005;**119**:1316-1330

[8] Baylis GC, Moore BO. Hippocampal lesions impair spatial response selection in the primate. Experimental Brain Research. 1994;**98**:110-118

[9] Parkinson JK, Murray EA, Mishkin M. A selective mnemonic role for the hippocampus in monkeys: Memory for the location of objects. Journal of Neuroscience. 1988;**8**:4159-4167

[10] Sutherland RJ, Rudy JW. Configural association theory: The role of the hippocampal-formation in learning, memory, and amnesia. Psychobiology. 1989;**17**:129-144

[11] Rudy JW, Sutherland RJ. Configural association theory and the hippocampal formation: An appraisal and reconfiguration. Hippocampus. 1995;**5**:375-389

[12] O'Reilly RC, Rudy JW. Computational principles of learning in the neocortex and hippocampus. Hippocampus. 2000;**10**:389-397

[13] O'Reilly RC, Rudy JW. Conjunctive representations in learning and memory: Principles of cortical and hippocampal function. Psychological Review. 2001;**108**:311-345

[14] Gray JA, McNaughton N. The Neuropsychology of Anxiety. 2nd ed. New York: Oxford University Press; 2000

[15] Davidson TL, Jarrard LE. The hippocampus and inhibitory learning: A gray area. Neuroscience & Biobehavioral Reviews. 2004;**28**:261-271

[16] Alvarado MC, Rudy JW. A comparison of kainic acid plus colchicine and ibotenic acid-induced hippocampal formation damage on four configural tasks in rats. Behavioral Neuroscience. 1995;**109**:1052-1062

[17] Rudy JW, Sutherland RJ. The hippocampal formation is necessary for rats to learn and remember configural discriminations. Behavioral Brain Research. 1989;**34**:97-109

[18] Sutherland R J, McDonald RJ. Hippocampus, amygdala, and memory deficits in rats. Behavioral Brain Research. 1990;**37**:57-79

[19] Davidson TL, McKernan MG, Jarrard LE. Hippocampal lesions do not impair negative patterning: A challenge to configural association theory. Behavioral Neuroscience. 1993;**107**:227-234

[20] Douglas RJ. The hippocampus and behavior. Psychological Bulletin. 1967;**67**:416-442

[21] Bland BH, Vanderwolf CH. Diencephalic and hippocampal mechanisms of motor activity in rat: Effects of posterior hypothalamic stimulation on behavior and hippocampal slow-wave activity. Brain Research. 1972;**43**:67-88

[22] Frederickson CJ, Whishaw IQ. Hippocampal EEG during learned and unlearned behavior in the rat. Physiology & Behavior. 1977;**18**:597-603

[23] Vanderwolf CH. Hippocampal electrical activity and voluntary movement in the rat. Electroencephalography and Clinical Neurophysiology. 1969;**26**:407-418

[24] Whishaw IQ, Vanderwolf CH. Hippocampal EEG and behavior: Changes in amplitude and frequency of RSA (theta rhythm) associated with spontaneous and learned movement patterns in rats and cats. Behavioral Biology. 1973;**8**:461-484

[25] Brankack J, Talnov A, Shin J, Matsumoto G. Task-related theta frequency changes in rats trained to perform an auditory discrimination. Abstracts—Society for Neuroscience. 1999;**25**:1386

[26] Gray JA. Sodium amobarbital, the hippocampal theta rhythm, and the partial reinforcement extinction effect. Psychological Review. 1970;**77**:465-480

[27] Gray JA, Ball GG. Frequency-specific relation between hippocampal theta rhythm, behavior, and amobarbital action. Science. 1970;**168**:1246-1248

[28] Kasicki S, Jeleń P, Olszewski M, Sławińska U. Electrical hippocampal activity during danger and safety signals in classical conditioning in the rat. Acta Neurobiologiae Experimentalis (Wars). 2009;**69**:119-128

[29] Masuoka T, Fujii Y, Kamei C. Participation of the hippocampal theta rhythm in memory formation for an eight-arm radial maze task in rats. Brain Research. 2006;**1103**:159-163

[30] Olvera-Cortés E, Cervantes M, González-Burgos I. Place-learning, but not cue-learning training, modifies the hippocampal theta rhythm in rats. Brain Research Bulletin. 2002;**58**:261-270

[31] Olvera-Cortés E, Cervantes M, González-Burgos I. Increase of the hippocampal theta activity in the Morris water maze reflects learning rather than motor activity. Brain Research Bulletin. 2004;**62**:379-384

[32] Sakimoto Y, Sakata S. The decline in rat hippocampal theta activity during response inhibition for the compound stimulus of negative patterning and simultaneous feature-negative tasks. Behavioral Brain Research. 2013;**257**:111-117

[33] Sakimoto Y, Sakata S. Change in hippocampal theta activity with transfer from simple discrimination tasks to a simultaneous feature-negative task. Frontiers in Behavioral Neuroscience. 2014;**8**:159

[34] Sakimoto Y, Sakata S. Hippocampal theta activity during behavioral inhibition for conflicting stimuli. Behavioral Brain Research. 2014;**275**:183-190

[35] Sakimoto Y, Sakata S. Behavioral inhibition during a conflict state elicits a transient decline in hippocampal theta power. Behavioral Brain Research. 2015;**290**:70-76

[36] Sakimoto Y, Sakata S. Change in hippocampal theta activity during behavioral inhibition for a stimulus having an overlapping element. Behavioral Brain Research. 2015;**282**: 111-116

[37] Sakimoto Y, Hattori M, Takeda K, Okada K, Sakata S. Hippocampal theta wave activity during configural and non-configural tasks in rats. Experimental Brain Research. 2013;**225**:177-185

[38] Sakimoto Y, Okada K, Takeda K, Sakata S. Transient decline in hippocampal theta activity during the acquisition process of the negative patterning task. PLoS One. 2013;**8**:e70756

[39] Sakimoto Y, Okada K, Hattori M, Takeda K, Sakata S. Neural activity in the hippocampus during conflict resolution. Behavioral Brain Research. 2013;**237**:1-6

[40] Sakimoto Y, Takeda K, Okada K, Hattori M, Sakata S. Transient decline in rats' hippocampal theta power relates to inhibitory stimulus-reward association. Behavioral Brain Research. 2013;**246**:132-138

[41] Sato N, Sakata S. Hippocampal theta activity during delayed nonmatching-to-sample performance in rats. Psychobiology. 1999;**27**:331-340

[42] Shin J, Lu BL, Talnov A, Matsumoto G, Brankack J. Reading auditory discrimination behaviour of freely moving rats from hippocampal EEG. Neurocomputing. 2001;**38-40**: 1557-1566

[43] Teitelbaum H, McFarland WL, Mattsson JL. Classical conditioning of hippocampal theta patterns in the rat. Journal of Comparative and Physiological Psychology. 1977;**91**: 674-681

[44] Yoder RM, Pang KC. Involvement of GABAergic and cholinergic medial septal neurons in hippocampal theta rhythm. Hippocampus. 2005;**15**:381-392

[45] Buzsáki G. Generation of hippocampal EEG patterns. In: Isaacson RL, Pribram KH, editors. The Hippocampus. Vol. 3. New York: Plenum Press; 1986. pp. 137-167

[46] Fox SE, Wolfson S, Ranck JB. Hippocampal theta-rhythm and the firing of neurons in walking and urethane anesthetized rats. Experimental Brain Research. 1986;**62**:495-508

[47] Mitchell SJ, Ranck JB. Generation of theta rhythm in medial entorhinal cortex of moving rats. Brain Research. 1980;**189**:49-66

[48] O'Keefe J. Hippocampal neurophysiology in the behaving animal. In: Andersen P, Morris R, Amaral D, Bliss T,O'Keefe J, editors. The Hippocampus Book. New York: Oxford University Press; 2008. pp. 475-548

[49] Grastyán E, Lissak K, Madarasz I, Donhoffer H. Hippocampal electrical activity during the development of conditioned reflexes. Electroencephalography and Clinical Neurophysiology. 1959;**11**:409-430

[50] Monmaur P, Breton P. Elicitation of hippocampal theta by intraseptal carbachol injection in freely moving rats. Brain Research. 1991;**544**:150-155

[51] Sun MK, Zhao WQ, Nelson TJ, Alkon DL. Theta rhythm of hippocampal CA1 neuron activity: Gating by GAB aergic synaptic depolarization. Journal of Neurophysiology. 2001;**85**:269-279

[52] Lee MG, Chrobak JJ, Sik A, Wiley RG, Buzsáki G. Hippocampal theta activity following selective lesion of the septal cholinergic system. Neuroscience. 1994;**62**:1033-1047

[53] Sakimoto Y, Mitsushima D. Rapid synaptic plasticity within 5 min: experienced episode increases diversity of excitatory and inhibitory synapses in the hippocampal CA1. Abstracts Journal of Physiological Science. 2016;**66**:s127

Chapter 3

Intraoperative Electroencephalography During Aortic Arch Surgery

Takashi Murashita

Additional information is available at the end of the chapter

http://dx.doi.org/10.5772/68024

Abstract

Since its introduction, deep hypothermic circulatory arrest has been widely used for cerebral protection during aortic arch surgery. The use of electroencephalogram plays an important role in intraoperative neurophysiologic monitoring. Systemic cooling to the point of electrocerebral inactivity has been thought to ensure optimal neuroprotection from the ischemic injury during circulatory arrest. Therefore, electroencephalogram can guide surgeons to induce deep hypothermic circulatory arrest at an optimal timing. In the meantime, along with the advent of adjunctive cerebral perfusion techniques, there is a certain trend that circulatory arrest is induced at higher degrees than traditional deep hypothermic approach, called moderate hypothermic circulatory arrest. The role of electroencephalogram in this approach has not been well established yet, but some studies suggested the importance of intraoperative electroencephalogram in this approach as well. Electroencephalogram is also utilized in emerging operative techniques called hybrid arch repair. To conclude, intraoperative use of electroencephalogram can greatly contribute to cerebral protection in the field of aortic arch surgery, and surgeons should be familiar with its mechanism, indication, and interpretation.

Keywords: circulatory arrest, neurophysiological monitoring, aortic arch surgery, hypothermia

1. Introduction

Since its introduction, deep hypothermic circulatory arrest has been widely utilized for cerebral protection during aortic arch operations [1]. By inducing hypothermia to the brain and visceral organs, tissue oxygen and metabolic demands are reduced to the extent that the period of ischemia resulting from circulatory arrest can be well withstood [2–5]. Because the brain is particularly sensitive to transient periods of hypoxia, cerebral protection is essential during aortic arch operations. Despite the advancement of surgical techniques, perioperative neurological complications following aortic arch operations are still reported to be as high as 5–8% in the current era [6–8]. Therefore, optimal methods how to induce circulatory arrest safely are still debated.

It has been shown that body temperature measurement is not a sufficient indicator of brain temperature [9]. Stone and colleagues reported that when profound hypothermia is rapidly induced and reversed, temperature measurements made at standard monitoring sites may not reflect cerebral temperature. Although a number of modalities, such as near-infrared spectroscopy and transcranial cerebral oximetry, have been introduced to monitor the brain during aortic arch operations, no single technique has proven to be a perfect monitoring tool. A method of physiological monitoring, intraoperative electroencephalography (EEG), was introduced by Ganzel and colleagues in 1997 [10]. The viewpoint is that maximal cerebral protection is achieved at temperatures sufficient to induce electrocerebral inactivity on EEG, under the assumption that maximal suppression of cerebral metabolic activity is achieved at electrocerebral inactivity [2, 11]. Stecker and colleagues reported that the process of cooling to electrocerebral inactivity produced a uniform degree of cerebral protection, independent of the actual nasopharyngeal temperature [12]. Consequently, many institutions have introduced intraoperative EEG to allow for the identification of electrocerebral inactivity before initiating circulatory arrest [13–16], which leads to average minimum temperatures of less than 16°C [17, 18].

2. The intraoperative use of EEG during aortic arch operations

2.1. EEG changes during systemic cooling and rewarming

Keenan and colleagues provided a review about neurological monitoring during aortic arch surgery [19]. EEG is monitored through the process of systemic cooling to provide occurring assessment of electrocerebral activity as a marker for the extent of hypothermia-mediated metabolic suppression in the brain. EEG monitoring is generally provided by using gold disc electrodes attached to the scalp according to the International System of Electrode Placement. Baseline EEG needs to be obtained after anesthetic induction but before initiation of cardiopulmonary bypass and systemic cooling. Baseline EEG gives us the identification of baseline asymmetry or other abnormal findings in electrocerebral activity. With the initiation of systemic cooling, EEG amplitude begins to diminish. The sensitivity of EEG should be increased for better assessment of low-amplitude activity. Because electrocerebral activity is significantly influenced by anesthesia, anesthetic agents are usually discontinued during the systemic cooling in order to mitigate the confounding impact these drugs have on interpreting EEG.

Stecker and colleagues reported the pattern of EEG electrocerebral activity during aortic arch operations requiring deep hypothermic circulatory arrest [12, 18]. Between a nasopharyngeal temperature of 21.5 and 34.2°C, a majority of patients are found to have either lateralized, generalized, or bilateral independent periodic discharges, or transient and synchronous

http://dx.doi.org/10.5772/68024

increases in EEG wave amplitude, against a background of continuous electrocerebral activity. Along with further systemic cooling, a gradual decrease in EEG continuity is found until the onset of a burst suppression pattern between 15.7 and 33.0°C. Finally, a progression to complete electrocerebral inactivity is found between 12.5 and 27.2°C (**Figure 1**).

During systemic rewarming, a reversed progression from electrocerebral inactivity back to normal amplitude continuous activity is found; however, the temperature points at which changes in EEG pattern tend to occur are different compared with the process of systemic

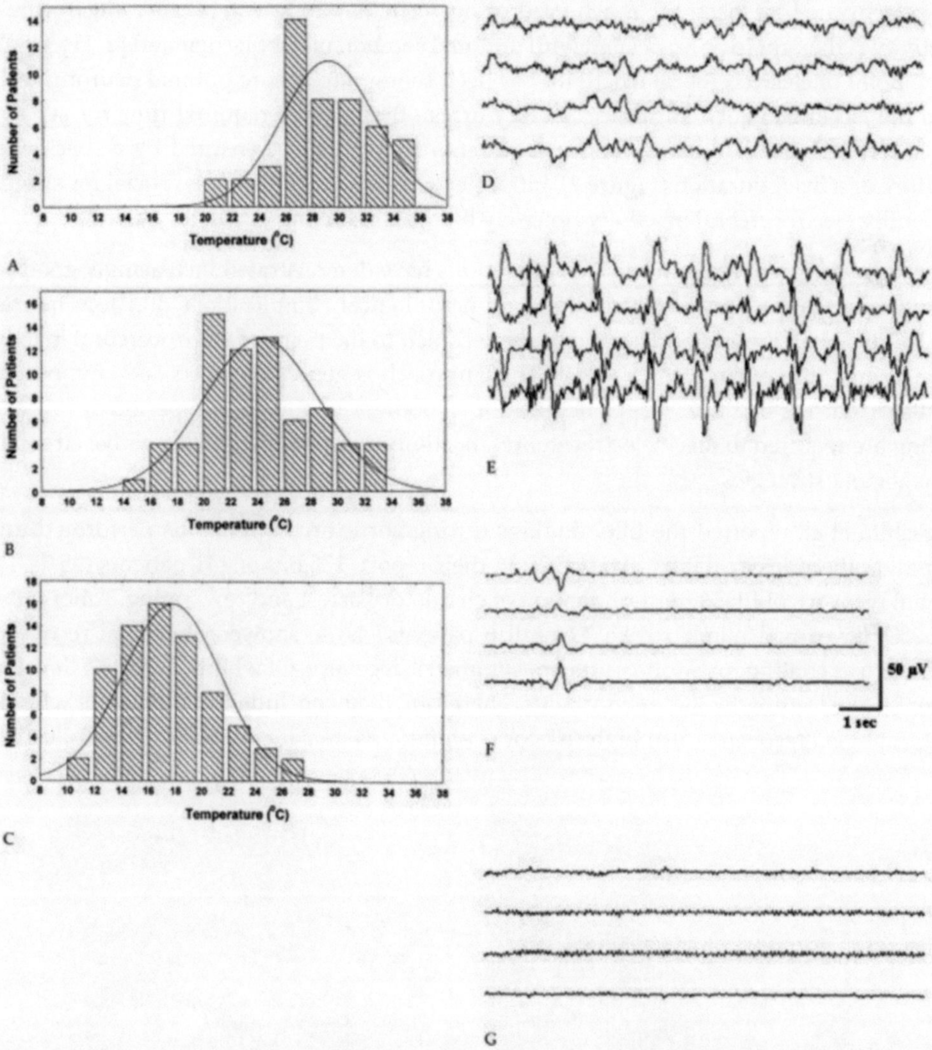

Figure 1. Distribution of nasopharyngeal temperatures at which various electroencephalogram landmarks occur: (A) appearance of periodic complexes, (B) appearance of burst suppression, and (C) electrocerebral inactivity. Examples of typical electroencephalogram patterns during systemic cooling are also shown: (D) precooling, (E) appearance of periodic complexes, (F) appearance of burst suppression, and (G) electrocerebral inactivity. Each of the electroencephalogram samples represent the following four channels from the left hemisphere (Fp1–F7, F7–T3, T3–T5, and T5–O1).

cooling. Therefore, while the changes of EEG pattern through the process of systemic cooling and rewarming are somewhat predictable, the temperature at which these EEG changes occur is significantly variable between patients. The required time for systemic cooling and rewarming is also variable and much depends of the patient and procedural factors [17, 18].

2.2. EEG findings during deep hypothermic circulatory arrest

The optimal degree of hypothermia before conducting circulatory arrest is still debated and remains a controversy in the field of aortic arch operations. Traditionally, the patient is cooled until electrocerebral inactivity is achieved prior to circulatory arrest, because electrocerebral inactivity is thought to be associated with minimal cerebral metabolic demand [2, 11]. Cooling to the point of electrocerebral inactivity has been thought to ensure optimal neuroprotection from the ischemic injury during circulatory arrest. Because the required time for achieving electrocerebral inactivity varies between patients and cannot be ensured by a specific temperature or a fixed duration (**Figure 2**), intraoperative EEG monitoring is crucial for surgeons to identify electrocerebral inactivity precisely before conducting circulatory arrest.

Previous reports regarding aortic arch operations have demonstrated increasingly good perioperative surgical outcomes, including low neurological complications and low mortality [13–16, 20–22]. In these reports, patients were cooled to the point of electrocerebral inactivity prior to initiation of circulatory arrest. This approach is generally called deep hypothermic circulatory arrest because deeper degrees of hypothermia and longer periods of systemic cooling are required to reach electrocerebral inactivity compared with alternative circulatory management strategies.

Murashita et al. reported the EEG findings during aortic arch operations performed under deep hypothermic circulatory arrest [22]. In their report, 135 out of 141 patients (95.7%) had normal recovery of EEG after termination of circulatory arrest and rewarming. Among them, 3 (2.2%) developed minor stroke. Overall 6 patients (4.3%) showed abnormal recovery of EEG, such as continuous suppression or asymmetric recovery. Of whom, 2 (33.3%) developed major stroke leading to 30-day mortality. Therefore, they concluded that patients who have abnormal EEG recovery are at high risk for postoperative major neurological complications.

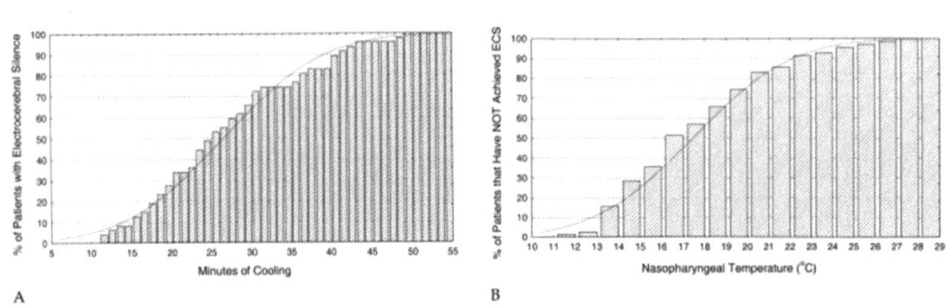

Figure 2. (A) The cumulative probability of electrocerebral silence on electroencephalogram as a function of cooling time. (B) The cumulative probability that electrocerebral silence is not achieved for temperatures above that indicated.

http://dx.doi.org/10.5772/68024

2.3. EEG findings during moderate hypothermic circulatory arrest

Although a deep hypothermic circulatory arrest is an established method in the field of aortic arch operations, there are concerns that the extremely low temperature can lead to adverse outcomes, such as hypothermia-related coagulopathy [23, 24], prolonged periods of cardiopulmonary bypass, or direct hypothermic neuronal injury [25–27]. In addition, the introduction of adjunctive cerebral perfusion techniques, such as retrograde and antegrade cerebral perfusion has allowed continued perfusion and cooling of the brain after systemic circulatory arrest. As a result, a number of institutions have been using a circulatory strategy of more moderate degrees of systemic hypothermia with adjunctive cerebral perfusion in aortic arch operations [28–33]. This circulatory strategy, generally called as moderate hypothermic circulatory arrest, has provided comparable or better surgical outcomes than traditional deep hypothermic circulatory arrest [34, 35]. There is a great deal of controversy regarding the superiority of moderate versus deep hypothermia. However, a trend of using higher degrees of hypothermic circulatory arrest than traditional deep hypothermia seems likely to continue.

Unlike deep hypothermic circulatory arrest, the role of intraoperative EEG monitoring in the setting of moderate hypothermic circulatory arrest has not well been established. In the setting of moderate hypothermic circulatory arrest, the timing of induction of circulatory arrest is usually determined based on the nasopharyngeal temperature between 20 and 28°C. There is a wide variety in the current literature regarding the definition of moderate hypothermia as well as the location of temperature measurement [36]. Most of the patients still demonstrate some form of electrocerebral activity in EEG at moderate temperatures, and the electrophysiological behavior of the brain around the time of circulatory arrest and the establishment of adjunctive cerebral perfusion techniques remains in the discovery phase.

Keenan and colleagues reported EEG findings during moderate hypothermic circulatory arrest during hemiarch replacement [37]. Their study included 71 patients who underwent hemiarch replacement with moderate hypothermic circulatory arrest. Nobody reached to electrocerebral inactivity at the time of circulatory arrest. Among 71 patients, 32 (45%) demonstrated an abrupt loss of electrocerebral activity immediately after circulatory arrest, indicative cerebral ischemia. However, the majority of them restored electrocerebral activity following establishment of unilateral antegrade cerebral perfusion (**Figure 3**). In the cases where unilateral selective antegrade cerebral perfusion (SACP) did not resolve the loss of electrocerebral activity, bilateral antegrade cerebral perfusion or further systemic cooling were required. **Figure 4** shows the case where unilateral selective antegrade cerebral perfusion did not restore the electrocerebral activity, but bilateral antegrade cerebral perfusion showed partial return of activity. In the remaining 39 patients (55%), electrocerebral activity was maintained in the brief period between circulatory arrest and selective antegrade cerebral perfusion and persisted after cerebral perfusion was established. There were no baseline characteristics between patients who had an abrupt loss of electrocerebral activity and those who did not. There were no postoperative stroke, transient ischemic attack, or permanent mental status change in both groups. They concluded that loss of electrocerebral activity following moderate hypothermic circulatory arrest can be restored by using adequate antegrade cerebral perfusion technique.

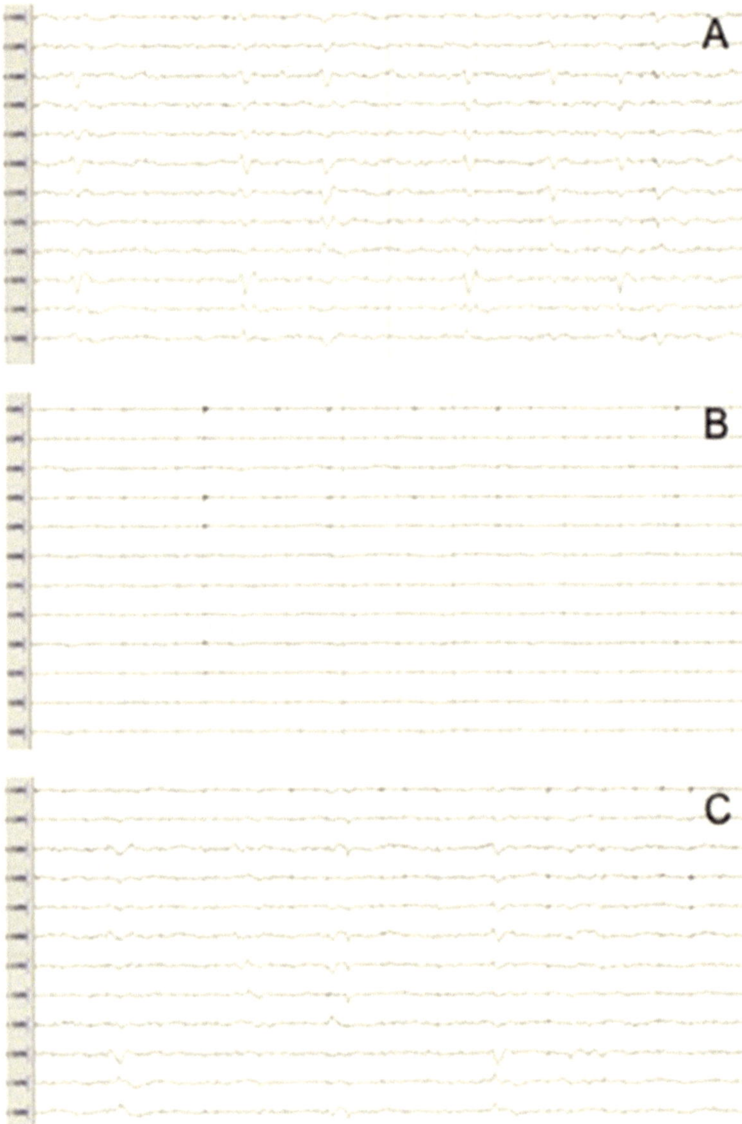

Figure 3. Restoration of electroencephalography (EEG) activity after circulatory arrest with the establishment of unilateral selective antegrade cerebral perfusion (SACP). (A) EEG before arrest shows burst suppression (nasopharyngeal temperature: 26.6°C). (B) EEG immediately after arrest demonstrates loss of electrocerebral activity. (C) EEG after unilateral SACP shows return of burst suppression pattern.

2.4. EEG use in emerging operative techniques

With the technical advancement of thoracic endovascular aortic repair, an innovative strategy that can avoid cardiopulmonary bypass and hypothermic circulatory arrest has emerged and become available for the surgical management of aortic arch pathology. That approach

http://dx.doi.org/10.5772/68024

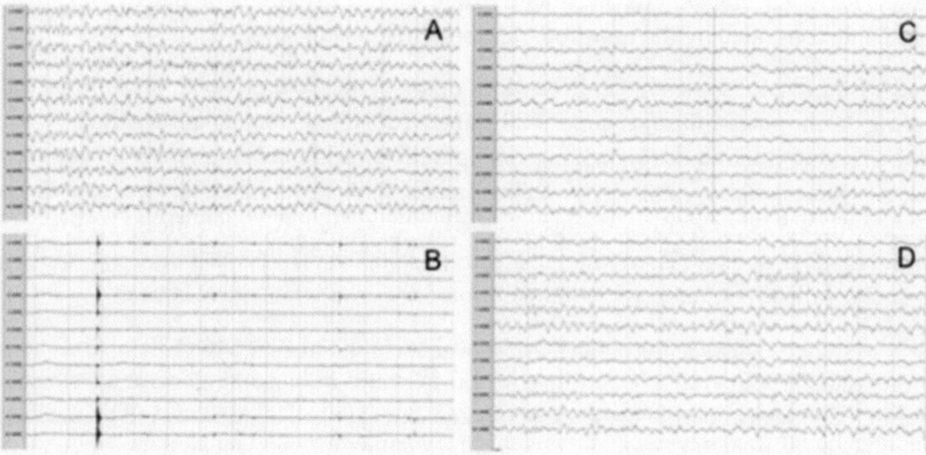

Figure 4. Persistent loss of left-sided electrocerebral activity after establishment of selective antegrade cerebral perfusion (SACP). (A) Electroencephalogram (EEG) before arrest shows a continuous pattern with diffuse wave slowing (nasopharyngeal temperature: 28.1°C). (B) EEG after arrest demonstrates loss of electrocerebral activity. (C) EEG after unilateral SACP shows persistent loss of electrocerebral activity in the left-sided leads. (D) EEG after transition to bilateral SACP shows partial return of left-sided activity.

is known as a "hybrid" arch repair [38–42]. This technique usually consists of two distinct operations. The first operation is the open procedure for extra-anatomical bypasses of the great arch vessels. When there is concomitant ascending or proximal arch pathology, they need to be replaced. In that case, cardiopulmonary bypass and circulatory arrest will be necessary in the same way with conventional aortic arch operations. Therefore, intraoperative neurological monitoring using EEG needs to be taken into consideration. However, when ascending aorta and proximal arch replacement are not required, the first operation of hybrid arch repair does not require cardiopulmonary bypass or circulatory arrest. In that case, EEG with or without near-infrared spectroscopy is often used to detect brain ischemia, especially when innominate artery or left common carotid artery is bypassed [42]. The second operation consists of deployment of a stent-graft in which the proximal landing zone is within the ascending aorta or proximal aortic arch and the distal landing zone is within the descending aorta. Because the great arch vessels are already bypassed in the first operation, there is little or no concern for brain ischemia during the second operation. The use of EEG in the second operation is not as important as the first operation, and the attention is usually more paid to spinal cord ischemia in the second operation.

2.5. The effect of anesthetic agents on EEG

Anesthetic agents have a significant impact on the findings of EEG [43]; therefore, the intraoperative management of anesthetic agents is very important during aortic arch operations. Generally, intravenous anesthetic drugs, such as propofol, benzodiazepines, and barbiturates can have significant effects on EEG findings. Even at small doses, these drugs can induce burst suppression in EEG, and these changes can mislead physicians during the process of systemic cooling. Therefore, intravenous anesthetic drugs should usually be avoided, and their use should be discontinued by the time body temperature gets to 30–33°C.

Instead of intravenous anesthetic drugs, low doses of the halogenated inhalational agents including desflurane, isoflurane, and sevoflurane with opioids for analgesia generally consist of the basis of anesthesia during aortic arch operations. Although high dose of these agents can lead to increasing suppression in electrocerebral activity, low dose of these agents generally required for adequate surgical anesthesia is not associated with significant effects on EEG amplitude and frequency. Therefore, detection of shifts toward lower frequency and amplitude can be indicative of cerebral ischemia or other neurophysiological issues.

Besides anesthetic agents, non-pharmacological factors can also affect EEG findings. For example, reduction in systemic and local blood pressure induced by general anesthesia can affect EEG. Changes in partial pressure of oxygen and carbon dioxide induced by mechanical ventilation can change EEG findings as a result of either altered oxygen delivery or changes in cerebral blood flow. Therefore, it is crucial to take these effects into consideration when interpreting EEG findings.

3. Conclusions

Aortic arch operation is one of the most complex surgeries in the current era. The introduction of hypothermic circulatory arrest has provided a great safety in protection of organs, especially for brain. Neurophysiological monitoring using intraoperative EEG plays a critical role in this field to help surgeons direct circulatory management and give clues about conditions of brain ischemia. With the traditional deep hypothermic circulatory arrest approach, electrocerebral inactivity is usually achieved, and surgeons are ensured that circulatory arrest can be achieved safely. However, with the advent of adjunctive cerebral perfusion techniques, there is a certain trend that hypothermic circulatory arrest is achieved at higher degrees (moderate hypothermic circulatory arrest) than deep hypothermia where electrocerebral inactivity is not achieved, and the surgical outcomes with moderate hypothermia have been favorable. Moderate hypothermic technique can avoid the issues associated with deep hypothermia, such as coagulopathy, prolonged cardiopulmonary bypass, and direct neuronal injury. The electrophysiological findings in moderate hypothermic circulatory arrest have not been established and need to be studied further. It is highly likely that neurological monitoring using EEG in the aortic arch operations will continue to be viewed as a crucial modality to ensure optimal patient safety and as the field continues to develop with new circulatory management strategies and operative techniques.

Author details

Takashi Murashita

Address all correspondence to: tmurashita@gmail.com

Heart and Vascular Institute, West Virginia University, Morgantown, WV, USA

References

[1] Griepp RB, Stinson EB, Hollingsworth JF, Buehler D. Prosthetic replacement of the aortic arch. J Thorac Cardiovasc Surg. 1975;**70**(6):1051-63

[2] Mezrow CK, Midulla PS, Sadeghi AM, Gandsas A, Wang W, Dapunt OE, Zappulla R, Griepp RB. Evaluation of cerebral metabolism and quantitative electroencephalography after hypothermic circulatory arrest and low-flow cardiopulmonary bypass at different temperatures. J Thorac Cardiovasc Surg 1994;**107**(4):1006-19

[3] McCullough JN, Zhang N, Reich DL, Juvonen TS, Klein JJ, Spielvogel D, Ergin MA, Griepp RB. Cerebral metabolic suppression during hypothermic circulatory arrest in humans. Ann Thorac Surg 1999;**67**(6):1895-9

[4] Strauch JT, Spielvogel D, Lauten A, Zhang N, Rinke S, Weisz D, Bodian CA, Griepp RB. Optimal temperature for selective cerebral perfusion. J Thorac Cardiovasc Surg. 2005;**130**(1):74-82

[5] Khaladj N, Peterss S, Pichlmaier M, Shrestha M, von Wasielewski R, Hoy L, Haverich A, Hagl C. The impact of deep and moderate body temperatures on end-organ function during hypothermic circulatory arrest. Eur J Cardiothorac Surg. 2011;**40**(6):1492-9

[6] Sundt TM, III., Orszulak TA, Cook DJ, Schaff HV. Improving results of open arch replacement. Ann Thorac Surg. 2008;**86**(3):787-96

[7] Okita Y, Miyata H, Motomura N, Takamoto S, Japan Cardiovascular Surgery Database Organization. A study of brain protection during total arch replacement comparing antegrade cerebral perfusion versus hypothermic circulatory arrest, with or without retrograde cerebral perfusion: analysis based on the Japan Adult Cardiovascular Surgery Database. J Thorac Cardiovasc Surg. 2015;**149**(2 Suppl):S65-73

[8] Kaneko T, Aranki SF, Neely RC, Yazdchi F, McGurk S, Leacche M, Shekar PS. Is there a need for adjunct cerebral protection in conjunction with deep hypothermic circulatory arrest during noncomplex hemiarch surgery? J Thorac Cardiovasc Surg. 2014;**148**(6):2911-7

[9] Stone JG, Young WL, Smith CR, Solomon RA, Wald A, Ostapkovich N, Shrebnick DB. Do standard monitoring sites reflect true brain temperature when profound hypothermia is rapidly induced and reversed? Anesthesiology 1995;**82**(2):344-51

[10] Ganzel BL, Edmonds HL, Jr., Pank JR, Goldsmith LJ. Neurophysiologic monitoring to assure delivery of retrograde cerebral perfusion. J Thorac Cardiovasc Surg. 1997;**113**(4):748-55

[11] Michenfelder JD, Milde JH. The effect of profound levels of hypothermia (below 14 degrees C) on canine cerebral metabolism. J Cereb Blood Flow Metab. 1992;**12**(5):877-80

[12] Stecker MM, Cheung AT, Pochettino A, Kent GP, Patterson T, Weiss SJ, Bavaria JE. Deep hypothermic circulatory arrest: II. Changes in electroencephalogram and evoked potentials during rewarming. Ann Thorac Surg 2001;**71**(1):22-8

[13] Coselli JS, Crawford ES, Beall AC, Jr., Mizrahi EM, Hess KR, Patel VM. Determination of brain temperatures for safe circulatory arrest during cardiovascular operation. Ann Thorac Surg 1988;**45**(6):638-42

[14] Bavaria JE, Pochettino A, Brinster DR, Gorman RC, McGarvey ML, Gorman JH, Escherich A, Gardner TJ. New paradigms and improved results for the surgical treatment of acute type A dissection. Ann Surg. 2001;**234**(3):336-42

[15] Svensson LG, Blackstone EH, Rajeswaran J, Sabik JF, III., Lytle BW, Gonzalez-Stawinski G, Varvitsiotis P, Banbury MK, McCarthy PM, Pettersson GB, Cosgrove DM. Does the arterial cannulation site for circulatory arrest influence stroke risk? Ann Thorac Surg. 2004;**78**(4):1274-84

[16] Lima B, Williams JB, Bhattacharya SD, Shah AA, Andersen N, Gaca JG, Hughes GC. Results of proximal arch replacement using deep hypothermia for circulatory arrest: is moderate hypothermia really justifiable? 77. 2011;**11**(1438-44)

[17] James ML, Andersen ND, Swaminathan M, Phillips-Bute B, Hanna JM, Smigla GR, Barfield ME, Bhattacharya SD, Williams JB, Gaca JG, Husain AM, Hughes GC. Predictors of electrocerebral inactivity with deep hypothermia. J Thorac Cardiovasc Surg. 2014;**147**(3):1002-7

[18] Stecker MM, Cheung AT, Pochettino A, Kent GP, Patterson T, Weiss SJ, Bavaria JE. Deep hypothermic circulatory arrest: I. Effects of cooling on electroencephalogram and evoked potentials. Ann Thorac Surg. 2001;**71**(1):14-21

[19] Keenan JE, Benrashid E, Kale E, Nicoara A, Husain AM, Hughes GC. Neurophysiological intraoperative monitoring during aortic arch surgery. Semin Cardiothorac Vasc Anesth. 2016;**20**(4):273-82

[20] Englum BR, Andersen ND, Husain AM, Mathew JP, Hughes GC. Degree of hypothermia in aortic arch surgery—optimal temperature for cerebral and spinal protection: deep hypothermia remains the gold standard in the absence of randomized data. Ann Cardiothorac Surg. 2013;**2**(2):184-93

[21] Svensson LG, Crawford ES, Hess KR, Coselli JS, Raskin S, Shenaq SA, Safi HJ. Deep hypothermia with circulatory arrest. Determinants of stroke and early mortality in 656 patients. J Thorac Cardiovasc Surg 1993;**106**(1):19-28

[22] Murashita T, Pochettino A. Intraoperative electroencephalogram-guided deep hypothermia plus antegrade and/or retrograde cerebral perfusion during aortic arch surgery. J Card Surg. 2016;**31**(4):216-9

[23] Livesay JJ, Cooley DA, Reul GJ, Walker WE, Frazier OH, Duncan JM, Ott DA. Resection of aortic arch aneurysms: a comparison of hypothermic techniques in 60 patients. Ann Thorac Surg 1983;**36**(1):19-28

[24] Leshnower BG, Myung RJ, Thourani VH, Halkos ME, Kilgo PD, Puskas JD, Chen EP. Hemiarch replacement at 28°C: an analysis of mild and moderate hypothermia in 500 patients. Ann Thorac Surg. 2012;**93**(6):1910-5

[25] Kumral E, Yüksel M, Büket S, Yagdi T, Atay Y, Güzelant A. Neurologic complications after deep hypothermic circulatory arrest: types, predictors, and timing. Tex Heart Inst J 2001;**28**(2):83-8

[26] Welz A, Pogarell O, Tatsch K, Schwarz J, Cryssagis K, Reichart B. Surgery of the thoracic aorta using deep hypothermic total circulatory arrest. Are there neurological consequences other than frank cerebral defects? Eur J Cardiothorac Surg. 1997;**11**(4):650-6

[27] Reich DL, Uysal S, Sliwinski M, Ergin MA, Kahn RA, Konstadt SN, McCullough J, Hibbard MR, Gordon WA, Griepp RB. Neuropsychologic outcome after deep hypothermic circulatory arrest in adults. J Thorac Cardiovasc Surg 1999;**117**(1):156-63

[28] Vallabhajosyula P, Jassar AS, Menon RS, Komlo C, Gutsche J, Desai ND, Hargrove WC, Bavaria JE, Szeto WY. Moderate versus deep hypothermic circulatory arrest for elective aortic transverse hemiarch reconstruction. Ann Thorac Surg. 2015;**99**(5):1511-7

[29] Kornilov IA, Sinelnikov YS, Soinov IA, Ponomarev DN, Kshanovskaya MS, Krivoshapkina AA, Gorbatykh AV, Omelchenko AY. Outcomes after aortic arch reconstruction for infants: deep hypothermic circulatory arrest versus moderate hypothermia with selective antegrade cerebral perfusion. Eur J Cardiothorac Surg 2015;**48**(3):45-50

[30] Kamiya H, Hagl C, Kropivnitskaya I, Böthig D, Kallenbach K, Khaladj N, Martens A, Haverich A, Karck M. The safety of moderate hypothermic lower body circulatory arrest with selective cerebral perfusion: a propensity score analysis. J Thorac Cardiovasc Surg. 2007;**133**(2):501-9

[31] Pacini D, Leone A, Di Marco L, Marsilli D, Sobaih F, Turci S, Masieri V, Di Bartolomeo R. Antegrade selective cerebral perfusion in thoracic aorta surgery: safety of moderate hypothermia. Eur J Cardiothorac Surg. 2007;**31**(4):618-22

[32] Urbanski PP, Lenos A, Bougioukakis P, Neophytou I, Zacher M, Diegeler A. Mild-to-moderate hypothermia in aortic arch surgery using circulatory arrest: a change of paradigm? Eur J Cardiothorac Surg 2012;**41**(1):185-91

[33] Zierer A, El-Sayed Ahmad A, Papadopoulos N, Moritz A, Diegeler A, Urbanski PP. Selective antegrade cerebral perfusion and mild (28°C-30°C) systemic hypothermic circulatory arrest for aortic arch replacement: results from 1002 patients. J Thorac Cardiovasc Surg 2012;**144**(5):1042-9

[34] Leshnower BG, Myung RJ, Kilgo PD, Vassiliades TA, Vega JD, Thourani VH, Puskas JD, Guyton RA, Chen EP. Moderate hypothermia and unilateral selective antegrade cerebral perfusion: a contemporary cerebral protection strategy for aortic arch surgery. Ann Thorac Surg. 2010;**90**(2):547-54

[35] Tsai JY, Pan W, Lemaire SA, Pisklak P, Lee VV, Bracey AW, Elayda MA, Preventza O, Price MD, Collard CD, Coselli JS. Moderate hypothermia during aortic arch surgery is associated with reduced risk of early mortality. J Thorac Cardiovasc Surg. 2013;**146**(3):662-7

[36] Yan TD, Bannon PG, Bavaria J, Coselli JS, Elefteriades JA, Griepp RB, Hughes GC, LeMaire SA, Kazui T, Kouchoukos NT, Misfeld M, Mohr FW, Oo A, Svensson LG,

Tian DH. Consensus on hypothermia in aortic arch surgery. Ann Cardiothorac Surg. 2013;**2**(2):163-8

[37] Keenan JE, Wang H, Ganapathi AM, Englum BR, Kale E, Mathew JP, Husain AM, Hughes GC. Electroencephalography during hemiarch replacement with moderate hypothermic circulatory arrest. Ann Thorac Surg. 2016;**101**(2):631-7

[38] Bavaria J, Vallabhajosyula P, Moeller P, Szeto W, Desai N, Pochettino A. Hybrid approaches in the treatment of aortic arch aneurysms: postoperative and midterm outcomes. J Thorac Cardiovasc Surg. 2013;**145**(3 Suppl):85-90

[39] Murashita T, Matsuda H, Domae K, Iba Y, Tanaka H, Sasaki H, Ogino H. Less invasive surgical treatment for aortic arch aneurysms in high-risk patients: a comparative study of hybrid thoracic endovascular aortic repair and conventional total arch replacement. J Thorac Cardiovasc Surg. 2012;**143**(5):1007-13

[40] Narita H, Komori K, Usui A, Yamamoto K, Banno H, Kodama A, Sugimoto M. Postoperative outcomes of hybrid repair in the treatment of aortic arch aneurysms. Ann Vasc Surg. 2016;**34**:55-61

[41] Preventza O, Garcia A, Cooley DA, Haywood-Watson RJ, Simpson K, Bakaeen FG, Cornwell LD, Omer S, de la Cruz KI, Price MD, Rosengart TK, LeMaire SA, Coselli JS. Total aortic arch replacement: A comparative study of zone 0 hybrid arch exclusion versus traditional open repair. J Thorac Cardiovasc Surg. 2015;**150**(6):1591-8

[42] Andersen ND, Williams JB, Hanna JM, Shah AA, McCann RL, Hughes GC. Results with an algorithmic approach to hybrid repair of the aortic arch. J Vasc Surg 2013;**57**(3):655-67

[43] Sloan TB. Anesthetic effects on electrophysiologic recordings. J Clin Neurophysiol. 1998; **15**(3):217-26

Chapter 4

Mathematical Foundation of Electroencephalography

Michael Doschoris and Foteini Kariotou

Additional information is available at the end of the chapter

http://dx.doi.org/10.5772/68021

Abstract

Electroencephalography (EEG) has evolved over the years to be one of the primary diagnostic technologies providing information concerning the dynamics of spontaneous and stimulated electrical brain activity. The core question of EEG is to acquire the precise location and strength of the sources inside the human brain by knowledge of an electrical potential measured on the scalp. But in what way is the source recovered? Leaving aside the biological mechanisms on the cellular level responsible for the recorded EEG signals, we pay attention to the mathematical aspects of the narrative. Our goal is to provide a brief and concise introduction of the mathematical terminology associated with the modality of EEG. We start from the very beginning, presenting step by step the mathematical formulation behind EEG in a simple and clear manner, keeping the mathematical notation to a minimum. Whilst we serve only the key relations for the described problems, we focus specifically on the limitations of each modelling approach. In this fashion, the reader can appreciate the beauty of the formulas presented and discover every single piece of information encoded within these formulas.

Keywords: EEG, mathematical analysis, forward problem, inverse problem, spherical conductor, ellipsoidal conductor

1. Introduction

The human brain is a remarkable and fascinating organ exhibiting a tremendous complexity. It makes us unique and defines who we are. In spite of our scientific and technological progression, we do not know the particulars of its operating, and as we delve into its secrets, various surprises emerge, for example, nearly 100 previously unidentified brain areas have been recently discovered [1]. Consisting of an inconceivable network of interconnected nerve cells and fibres, continuously transporting and processing information, the brain is extremely vulnerable and requires paramount protection. Several layers of safety are incorporated starting with three connective sheets of tissue, called the meninges containing the cerebrospinal fluid, followed by plates of bones, the skull. Above machinery safeguards the brain from mechanical damage. On the other hand, a sophisticated barrier within the brain provides a natural defence against toxic or infective agents.

Examining the anatomy of the brain, we recognize three distinct regions. The largest part of the brain is the cerebrum, divided into two hemispheres. The outermost layer of the cerebrum is the cerebral cortex, consisting of four lobes. Cognitive awareness has its origins here. The second largest part of the brain is the cerebellum, located underneath the cerebrum and responsible for motor control and learning. Last, but not least, an integral part of the brain connecting the cerebrum with the spinal cord is the brainstem, regulating reflexes and crucial, basic life functions. Detailed information can be found in [2].

The operational status of the brain is based on an alternating chain of electrical and chemical events. On the microscopic level, encoding and transmitting of information via electrochemical signals is achieved by the active participation of neuronal and non-neuronal constituents. Brain cells communicate through synaptic transmissions by controlling chemical transmitters or ionic currents which flow across their membranes. As a consequence, an electromagnetic field is generated. For a far-reaching introduction on the subject, see [3]. The question at hand now lies in the possibility exploiting these provoked fields. It seems only reasonable that if a substantial number of cells form a critical mass, which activates synchronously, the emerging electric and magnetic field should be detectable. This is indeed the case and a deeply rooted concept in electrophysiology [4, 5]. From an electrofunctional point of view, the ionic micro-flow within a single brain cell creates an opposite polarity between two point electrical charges very close together, leading to the notion of a dipole, a physical quantity one could say consists of the 'fundamental unit', which produces the observed fields. Dipoles are characterized by a vector called moment, the product of the charge and distance, visualized as an arrow pointing from a minimum (negative charge) to a maximum (positive charge), ergo featuring direction and magnitude. For that reason, it may be argued that the macroscopic description of the brain's activity is best achieved when simulated as an array of dipoles, that is, a non-uniform distribution of positive and negative charges. According to the latter, if a small neighbourhood is stimulated, an excessive number of dipoles concur and their electric fields would add or cancel one another depending on the direction. This complicated and difficult situation can be avoided by introducing the concept of the equivalent current dipole (ECD), namely a single dipole which generates the identical electric field as all of the individual dipoles together, hence summarizing the net effect of all microscopic currents located in the distinct region of the brain under consideration. This is a widely used approximation concept in the framework of neuroelectromagnetism [6, 7]. On the other hand, when the exertion is no longer confined to a focal region of the brain, then every one of these regions is simulated by an equivalent current dipole, leading to a distribution of sources.

The main task and problem is to correlate active regions with associated generated electric fields. This essential step is closely connected with the installation of physical structures, namely a boundary or number of boundaries enclosing distinct regions with specific physical characteristics, such as conductivity. The head model obtained is termed the volume conductor model. Clearly, the level of details incorporated into the head model provides an analogous degree of operational freedom when it comes to investigate how the fields generated by brain cells are transmitted through various biological tissues towards measurement apparatus. As a result, the volume conductor model consists of the physical foundation for source analysis, which is categorized into two major problems. The first one is associated with the calculation of the electric potential, generated by known electrochemical sources within the brain, at precise points at the surface of the scalp. This is the forward electroencephalographic (EEG) problem [8–11]. The forward EEG problem has been extensively scrutinized for over 60 years since Wilson and Bayley [12] attempted to quantify the interplay between neuronal activity and the potentials they generate at the scalp. The reconstruction of the sources responsible for the recorded values is called the corresponding inverse EEG problem [13].

As of today, a high level of details can only be achieved with the aid of numerical models, which are generally categorized into boundary element models (BEMs) and finite element models (FEMs). Whereas boundary element models are adequate to portray major tissue compartments, such as the cerebrum and skull, they fail to represent detailed anatomical information within the compartments, such as the cerebral folding [14, 15]. Finite element models, on the other hand, are efficient in capturing these details, but are labour intensive and computationally demanding [16, 17].

Nonetheless, in order to gain a deeper comprehension of the problem a rigorous mathematical analysis is essential in providing a vital step towards the recognition of the underlying phenomena as well as identifying the limitations of the developed algorithms. The importance of mathematical analysis cannot be emphasized enough, since (i) it allows testing the impact of modifications regarding various variables upon the output of the system and provides further insight into underlying physical behaviour. (ii) It serves as validation tools for the numerical models.

2. The mathematical formulation of the EEG problem

Think of the following scenario. You are conducting a series of experimental or clinical studies, but out of curiosity and foremost for a better understanding of the way the system under consideration behaves, you desire to build a mathematical model interpreting as best as possible your measurements. But, where to begin? First of all, we will need a framework which is capable of, at least to a degree, explaining what happens and why. If such a framework does not exist, we have to formulate one. As mentioned in Section 1, the electrochemical activity of brain cells results in bioelectric sources which generate an electric field in the neighbourhood of the cells. This field varies generally in time. Consequently, electromagnetic phenomena materialize. Luckily for us, the framework interpreting this kind of phenomena and the starting point of our endeavour are Maxwell's equations, a set of four equations, namely

$$\nabla \times \mathbf{E} + \frac{\partial \mathbf{B}}{\partial t} = \mathbf{0},\ \nabla \times \mathbf{H} = \mathbf{J} + \frac{\partial \mathbf{D}}{\partial t},\ \nabla \cdot \mathbf{B} = 0,\ \nabla \cdot \mathbf{D} = \rho. \tag{1}$$

The inverted delta present in Eq. (1) is called del, or nabla, and consist of a mathematical device named operator, a symbol indicating that an action must be performed on what follows. The algebraic operations of the dot ($\cdot$) and cross ($\times$) product between two vectors should not be confused with the elementary operation of multiplication.

Maxwell equations connect the electric fields **E** and **B**, the displacement field **D** and the magnetizing field **H** with their sources, that is, the charge density ρ and the current density **J**. These fields have direction and magnitude and must be represented as vector functions (bold capital letters). The above set of equations is also called Maxwell's macroscopic equations, and in order to apply them, a relation between **D** and **E**, as well as **H** and **B** must be specified. For materials without polarization and magnetization, they are

$$\mathbf{D} = \varepsilon \mathbf{E},\ \mathbf{H} = \mu^{-1}\mathbf{B}, \tag{2}$$

where ε is the relative permittivity (dielectric constant) of the material and shows how strong the material influences the electric field **E**. Similarly, μ is the magnetic permeability of the material and provides the corresponding influence on the magnetic field **B**.

In what follows, we shall consider the following instance. For a finite medium, we introduce the characteristic dimension R, namely the smallest sphere with radius R which envelopes the medium under consideration. On the special occasion where the wavelength λ of the wave generated by the electromagnetic field is much larger than the characteristic dimension of the medium, namely $\lambda >> R$, then the corresponding time rates of change are very small. The latter observation leads to the quasi-static theory of Maxwell's equations, which take the form (see Ref. [18] for details)

$$\nabla \times \mathbf{E} = \mathbf{0},\ \nabla \times \mathbf{H} = \mathbf{J},\ \nabla \cdot \mathbf{B} = 0,\ \nabla \cdot \mathbf{D} = \rho. \tag{3}$$

It can be shown [19] that for a medium the size of the brain, R equals about 20 cm whereas λ about 400 m. Therefore, $\lambda \cong 2000\text{R}$ and the application of Maxwell's quasi-static equations are justified. Replacing Eq. (2) into Eq. (3), we immediately find

$$\nabla \times \mathbf{E} = \mathbf{0},\ \nabla \times \mathbf{B} = \mu\mathbf{J},\ \nabla \cdot \mathbf{B} = 0,\ \nabla \cdot \mathbf{D} = \varepsilon^{-1}\rho. \tag{4}$$

Let us focus on the first of Eq. (4). The curl of a vector, that is, the operator $\nabla\times$ captures the idea of how the vector field is circulating around a central axis. From this point of view, the electric field is irrotational. By a well-known identity of vector calculus, if the curl of a vector is zero, then the vector field in question can always be expressed as the gradient of a scalar field. In our case, the absence of circulation of **E** is caused by a continuously decreasing electric potential U along the direction of the electric field,

$$\mathbf{E} = -\nabla U. \tag{5}$$

An inherit characteristic of linear systems is the principle of superposition. Here, the electric fields are superposable, meaning that the electric field generated by a number of charges can be expressed as the vector sum of the electric fields generated by each charge separately; it

http://dx.doi.org/10.5772/68021

follows from Eq. (5) that the electric potentials are superposable as well. As a result, it is much easier to compute the provoked potential than the corresponding electric field.

At this point, the benefit of Eq. (5) is not clear yet. To show the usefulness of Eq. (5), we utilize a theorem of vector calculus stating that the divergence of the curl of a vector always vanishes. Applying the latter on the second of Eq. (3), we are left with

$$\nabla \cdot \mathbf{J} = 0 \tag{6}$$

and the current density of **J** is said to be solenoidal and expresses the steady-state condition that the charge density ρ is not changing in time. Moreover, an applied field in a resistive material, such as the brain, will induce a current of density $\mathbf{J}_i$, directly proportional to the applied field provided by a generalization of Ohm's law due to Kirchhoff (for a detailed historical analysis see Ref. [20]),

$$\mathbf{J}_{\mathrm{i}} = \sigma \mathbf{E}, \tag{7}$$

where the proportionality constant σ is termed conductivity. Therefore, the total current is given as the sum of the primary current $\mathbf{J}_{\mathrm{p}}$ responsible for the electric field, and the induced, or secondary, current $\mathbf{J}_{\mathrm{i}}$ given by Eq. (8), as

$$\mathbf{J} = \mathbf{J}_{\mathrm{p}} + \sigma \mathbf{E}. \tag{8}$$

Combining Eqs. (5), (6) and (8), we arrive at the following differential equation:

$$\nabla \cdot (\sigma\, \nabla U) = \nabla \cdot \mathbf{J}_{\mathrm{p}}, \tag{9}$$

which must be satisfied by the electric potential U. Simple vector calculus shows that

$$\nabla \cdot (\sigma\, \nabla U) = \nabla U \cdot \nabla \sigma + \sigma\, \Delta U, \tag{10}$$

where the symbol Δ is called the Laplacian operator.

When the conductivity varies in space, that is, the medium under consideration is inhomogeneous, consisting of compartments which are not of the same material, the quantity $\nabla\sigma$ differs from zero. On the other hand, when the conductivity is constant in space (homogeneous, isotropic) or constant by direction (anisotropic), then the gradient of σ vanishes. In the latter case, the electric potential U is related to the primary current $\mathbf{J}_{\mathrm{p}}$ by Poisson's equation,

$$\Delta U = \sigma^{-1} \nabla \cdot \mathbf{J}_{\mathrm{p}}. \tag{11}$$

Note that the 'source term', namely the right-hand side of Eq. (11), is not the primary current per se, but the divergence of $\mathbf{J}_{\mathrm{p}}$ multiplied by 1 over σ. In view of Eq. (11), the forward EEG problem is now formulated as follows. Given the primary current density $\mathbf{J}_{\mathrm{p}}$, calculate the electric potential on the surface of the medium. On the contrary, calculating $\mathbf{J}_p$ from the knowledge of U consists of the inverse EEG problem.

So far, we managed to derive at an equation which allows the computation of the electric potential, but our framework is still incomplete. Because the medium under consideration is finite, that is, confined in space by a closed surface-boundary, we need a set of additional constrains, the so-called boundary conditions, which U has to satisfy as well. This stems from the fact that when the medium, in which a wave propagates, displays alterations in its material properties (e.g. different conductivities), the wave is reflected, transmitted or both. In any case, at the interface S separating two regions V_1, V_2 the wave and its normal derivative must be continuous at the interface, namely

$$U_1 = U_2 \text{ and } \sigma_1 \partial_\nu U_1 = \sigma_2 \partial_\nu U_2 \text{ on } S, \tag{12}$$

where σ_1, σ_2 denote the conductivity of V_1, V_2, respectively. Above relations are valid only if no charged layer near the interface exists, that is, the absence of primary currents in the vicinity of S is secured. The first of Eq. (12) is known as Dirichlet condition, whereas the second is called a Neumann condition. Neumann conditions must be supplemented by the compatibility condition that the sum of all contributions of the normal derivative $\partial_\nu U$ on S must cancel out, that is,

$$\oint \partial_\nu U \, \mathrm{d}S = 0. \tag{13}$$

Further note that, by virtue of $\partial_\nu U$, the value of the electric potential is non-unique up to an additive constant. Eqs. (11)–(13) are all we need in order to tackle the EEG problem.

3. The brain modelled as a volume conductor

The next step in our journey is to introduce a geometry simple enough in order to carry out the mathematics associated, still adequate realistic in order to illustrate what happens. In practice, above specifications are never met. On the grounds that our interest is focused on the derivation of analytic formulas which will allow us to identify and recognize underlying phenomena, we restrain ourselves to the study of two particular geometries: (i) the spherical and (ii) the ellipsoidal. The distinctness of these two geometries lies in a different representation of the same point in space.

Before we continue, we have to incorporate our assumption regarding the nature of the primary current. For a single, localized dipole at point $\mathbf{r}_0$ and moment $\mathbf{Q}$, we consider

$$\mathbf{J}_p = \mathbf{Q}\delta(\mathbf{r} - \mathbf{r}_0), \tag{14}$$

where the functional $\delta(\mathbf{r})$ is the Dirac measure, a concept included in the analysis by the reason of representing the concentration of $\mathbf{J}_p$ to a single point [21].

3.1. The homogeneous spherical brain

The simplest possible geometry in order to represent the brain consists of a spherical homogeneous conductor with radius a and conductivity σ. The reasons are twofold. First of all, it is a good fit to the actual brain. Secondly, it is the only geometry for which data regarding geometrical and physical aspects are immediately available. For example, knowledge of the brains volume can be directly translated into the radius of the corresponding sphere. Importantly, the spherical geometry allows the deduction of explicit expressions for the quantities involved and therefore permits a thorough investigation of the behaviour of the system under scrutiny without running every single time a—mostly time-consuming—computer simulation.

The obvious question at hand is what part of the brain do we model? Clearly, since our initial model is based on homogeneity, it cannot represent the brain en masse, as mentioned in Section 1. So we start with the uppermost region of the human central nervous system, the cerebrum. It is therefore of uttermost importance to be aware of the strengths and weaknesses of the proposed model(s). Without any doubt, the homogeneous spherical model presents an unrealistic assumption of the brain-head system. So why should we bother with theoretical models? The answer is relatively simple. We need them in order to be able to draw conclusions when we move to build models of higher complexity. They serve the important task revealing gaps between forthcoming models, but more substantially they allow us to test the reliability of the introduced algorithms in a straightforward and timely matter. On the other hand, with the homogeneous model, activity in subcortical structures is impossible to detect. Moreover, the influence of the bone architecture enclosing the brain cannot be assessed as well. The latter implies that we actually do not record EEG data at all, but rather monitor the electrophysiology of the (exposed) cerebrum by electrocorticography (ECoG), or intracranial electroencephalography (iEEG).

3.1.1. Forward and inverse problem for a single dipole and multiple dipoles

Having aforementioned remarks in mind, let's begin finding a relation which connects the electric potential on the surface of our conductor model with the electric activity of cells inside. Our goal is achieved solving Eq. (11) combined with expression (14), namely

$$\Delta U = \sigma^{-1}\mathbf{Q} \cdot \nabla\delta(\mathbf{r} - \mathbf{r}_0), \; r < a \tag{15}$$

supplemented by the condition

$$\partial_r U = 0, \; r = a, \tag{16}$$

which follows at once from the second condition (12) expressing the circumstance that the conductivity outside the brain vanishes, and expresses the 'reality' that no current exists outside the brain.[1] Note that the compatibility condition (13) is automatically satisfied by Eq. (16).

[1]This statement is true only for a homogeneous conductor as described in Section 3.

Employing analytic techniques, the interested reader can find all details in Ref. [22]; it is not hard to show that the solution regarding Eqs. (15) and (16) evaluated at the surface is given as

$$U_{\text{surf}}(\hat{\mathbf{r}}, \mathbf{r}_0) = \sum_{n=1}^{\infty} \sum_{m=-n}^{n} \mathrm{A}_n^m(\hat{\mathbf{r}}_0) Y_n^m(\hat{\mathbf{r}}),\ \mathrm{A}_n^m(\hat{\mathbf{r}}_0) = \frac{\mathbf{Q}}{\sigma n a^{n+1}} \cdot \nabla_{\mathbf{r}_0}\left(r_0^n \overline{Y}_n^m(\hat{\mathbf{r}}_0)\right), \tag{17}$$

where a hat '^' denotes the unit vector, Y and $\overline{Y}$ are the spherical harmonics and corresponding complex conjugate, respectively, which are complex-valued functions, the latter with equal real part and imaginary part equal in magnitude but opposite in sign [23].

Expression (17) can be simplified using a summation formula [22, 24] yielding the following closed form:

$$U_{\text{surf}}(\hat{\mathbf{r}}, \mathbf{r}_0) = \frac{\mathbf{Q}}{4\pi\sigma} \cdot \left(2\frac{\mathbf{R}}{R^3} + \frac{1}{aR}\frac{R\hat{\mathbf{r}} + \mathbf{R}}{R + \hat{\mathbf{r}} \cdot \mathbf{R}}\right),\ \mathbf{R} = a\hat{\mathbf{r}} - \mathbf{r}_0, \tag{18}$$

where an italics-type capital letter denotes the magnitude of the corresponding vector. Eq. (18) consists of the simplest, straightforward expression regarding EEG data.

Let us now concentrate on the most important aspect when it comes to imaging modalities, such as EEG, namely the problem of identifying the primary source by means of a generated electromagnetic field. We recall that the notion of an equivalent dipole source has been adopted in order to summarize the entire microscopic currents located in the vicinity of a specific area in the brain. Notwithstanding, there does not exist an exclusive source configuration for each set of electroencephalographic measurements, constituting the corresponding inverse problem non-unique. The only way to eliminate non-uniqueness is to provide supplementary information's, that is, imposing additional assumptions. By introducing the assumption of an equivalent dipole source, the inverse problem can be solved exactly as we will show in the sequel.

The inverse problem for the homogeneous spherical conductor is formulated as follows. From surface measurements, we identify the potential U_{surf}, given via Eq. (17), from which we have to calculate the position $\mathbf{r}_0 = (x_0, y_0, z_0)$ and moment $\mathbf{Q} = (Q_x, Q_y, Q_z)$ of the dipole. In our case, this is easily achieved, noting that every detail regarding $\mathbf{r}_0$ and $\mathbf{Q}$ are encoded into the coefficients A_n^m of expansion (17). The latter are evaluated with the aid of the orthogonality condition

$$\mathrm{A}_n^m = \oint U_{\text{surf}}(\hat{\mathbf{r}}, \mathbf{r}_0) \overline{Y}_n^m(\hat{\mathbf{r}}) dS(\hat{\mathbf{r}}). \tag{19}$$

A word of caution with respect to Eq. (19). In order to employ the latter, we must know the surface potential on every single point via Eq. (17). In practice, the function U_{surf} is acquired as a continuous function via interpolation of the discrete set of EEG measurements. Moreover, since each vector contains three coordinates, in order to pinpoint the position and moment of the dipole, at least six equations are required. As a result, we expand the coefficients A_n^m, given via Eq. (17), for n = 1 and 2 providing eight relations in total. The first three of them, namely A_1^{-1}, A_1^0 and A_1^1, are proportional to the components Q_x, Q_y, Q_z of $\mathbf{Q}$, and thus we find

$$Q_x = \sqrt{\frac{2\pi}{3}}\sigma a^2 (A_1^{-1} - A_1^1),\ Q_y = -i\sqrt{\frac{2\pi}{3}}\sigma a^2 (A_1^{-1} + A_1^1),\ Q_z = 2\sqrt{\frac{\pi}{3}}\sigma a^2 A_1^0, \tag{20}$$

where i denotes the imaginary unit. It is not hard to show that whereas the difference $A_1^{-1} - A_1^1$ is real valued, the corresponding sum is imaginary. The remaining five coefficients A_2^m for $n = 2$ and $m = -2,-1,0,1,2$ connect the position of the dipole with its moments which are eliminated utilizing Eq. (20). Again, after some algebra we have

$$x_0 = \frac{a}{\sqrt{5}}\left(\frac{A_2^{-2}}{A_1^{-1}} - \frac{A_2^2}{A_1^1}\right),\ y_0 = -i\frac{a}{\sqrt{5}}\left(\frac{A_2^{-2}}{A_1^{-1}} + \frac{A_2^2}{A_1^1}\right),\ z_0 = \frac{2a}{\sqrt{5}}\frac{1}{A_1^1}\left(A_2^1 - \frac{A_2^2 A_1^0}{\sqrt{2}A_1^1}\right). \tag{21}$$

There is another way to recover the solution to the inverse problem for a single dipole. The approach illustrated provides a glimpse into the beauty of mathematical analysis. We will show that the uniqueness of the inverse problem, for a single dipole, is closely connected with the condition of attaining certain relations connecting the measured data. Considering that we need six equations to identify the source, we expand Eq. (17) for $n = 1$, 2 and express the resulting relation in Cartesian coordinates, yielding [25]

$$U_{\text{surf}} = A_1 x_0 + A_2 y_0 + A_3 z_0 + B_1 x_0{}^2 + B_2 y_0{}^2 + B_3 z_0{}^2 + C_1 x_0 y_0 + C_2 x_0 z_0 + C_3 y_0 z_0 + \ldots \tag{22}$$

where again the first three coefficients A_1, A_2, A_3 are proportional to the components of $\mathbf{Q}$, whereas the remaining six involve products of the position of the dipole with its moments. Substituting the expressions for Q_x, Q_y, Q_z into $B_1, B_2, B_3, C_1, C_2, C_3$, we arrive at the linear but overdetermined system $\mathbb{A}\mathbf{r}_0 = \mathbf{b}$ with

$$\mathbb{A} = \begin{bmatrix} 2A_1 & -A_2 & -A_3 \\ -A_1 & 2A_2 & -A_3 \\ -A_1 & -A_2 & 2A_3 \\ A_2 & A_1 & 0 \\ A_3 & 0 & A_1 \\ 0 & A_3 & A_2 \end{bmatrix},\ \mathbf{b} = \frac{6a^2}{5}\begin{bmatrix} B_1 \\ B_2 \\ B_3 \\ C_1 \\ C_2 \\ C_3 \end{bmatrix}. \tag{23}$$

Adopting Gauss elimination, we find that the position of the dipole is given as

$$x_0 = \frac{3}{2A_1A_2A_3}(A_1\mathbf{a}) \cdot \tilde{\mathbf{b}},\ y_0 = \frac{3}{2A_1A_2A_3}(A_2\mathbf{a}) \cdot (\mathbb{R}_z\tilde{\mathbf{b}}),\ y_0 = \frac{3}{2A_1A_2A_3}(A_3\mathbf{a}) \cdot (\mathbb{R}_y\tilde{\mathbf{b}}), \tag{24}$$

where $\mathbf{a} = (A_1, A_2, A_3)^T$, $\tilde{\mathbf{b}} = (-C_1, C_2, C_3)^T$, T denotes transposition, whereas $\mathbb{R}_y$ and $\mathbb{R}_z$ denote the rotation matrices about the y- and z-axis, respectively.

A very important aspect, revealed by the above analysis, is evidence that the dipoles position specified by relations (24) is unique, only if the recorded values for the coefficients present in Eq. (22) satisfy the following relations:

$$\begin{aligned} &2A_1A_2A_3(B_3 - B_2) + 3A_1(A_2^2 - A_3^2)C_2 + 3(A_2^2 + A_3^2)(A_3C_1 - A_2C_3) = 0, \\ &-2A_1A_2A_3(B_2 + 2B_3) + 3A_2(A_3^2 - A_1^2)C_3 + 3(A_1^2 + A_3^2)(A_1C_2 - A_3C_1) = 0. \end{aligned} \tag{25}$$

Indeed, replacing the analytic expressions for the coefficients of Eq. (22) into Eq. (25), these are trivially satisfied. Complementary, we briefly state that the least-square solution to Eq. (23) is $\mathbf{r}_0 = \mathbb{A}^+\mathbf{b} + (\mathbb{I} - \mathbb{A}^+\mathbb{A})\mathbf{Y}$, where $\mathbb{A}^+$ is the pseudoinverse of $\mathbb{A}$, and $\mathbf{Y}$ is an arbitrary vector [26]. Evidently, the presence of $\mathbf{Y}$ constitutes the solution non-unique. Nonetheless, for a single dipole the equality $\mathbb{A}^+\mathbb{A} = \mathbb{I}$ holds true, leading to the minimum norm solution $\mathbf{r}_0 = \mathbb{A}^+\mathbf{b}$.

However, in 'reality' EEG recordings provide values for the coefficients A_n^m, as given in Eq. (17), but our uniqueness criteria (25) are bound to the coefficients of Eq. (22). So, how do the correlate? Well, connection formulas are established by expanding the first eight terms of the surface potential (17). These terms are then rearranged to form an expression similar to Eq. (22). Comparing the resulting relation with Eq. (22) provides the connection. The final formulas can be found in Ref. [25].

Until now, we attended the situation where brain activity is simulated by a single dipole. What happens if a larger area or multiple areas are stimulated? Is a single dipole adequate to describe the event? Multiple areas of the cortex are often expected to be active at the same time, so the answer must be no. In Section 2, we mentioned that the electric fields are superposable. As a consequence, the surface potential due to N-dipole sources is computed as the sum of the potentials U_j corresponding to dipoles located at $\mathbf{r}_{0j}$ and strength $\mathbf{Q}_j$. In order to calculate their positions and moments, we require at least $6N$ equations. Although an analytic inversion algorithm can be derived [27], the steps necessary are cumbersome and beyond the scope of the present document. It remains, however, the question of identifying multiple localized sources by means of a recorded potential. Is it possible to be led into an erroneous conclusion when we have to recognize the number of activated areas? This situation can occur when data (coefficients) received are falsely interpreted as evoked by a single dipole. To show this, we first rewrite the surface potential resulting from a single dipole (17), in the form

$$U_{\text{surf}}(\hat{\mathbf{r}}, \mathbf{r}_0) = \frac{1}{4\pi\sigma} \sum_{n=1}^{\infty} \frac{2n+1}{na^{n+1}} (\mathbf{Q} \cdot \nabla_{\mathbf{r}_0}) r_0^n P_n(\hat{\mathbf{r}} \cdot \hat{\mathbf{r}}_0), \tag{26}$$

employing the addition theorem for Legendre functions P_n. The same expression is valid on the occasion of N dipoles, that is,

$$U_{\text{surf}}(\hat{\mathbf{r}}, \mathbf{r}_{0j}) = \frac{1}{4\pi\sigma} \sum_{n=1}^{\infty} \frac{2n+1}{na^{n+1}} \sum_{j=1}^{N} (\mathbf{Q}_j \cdot \nabla_{\mathbf{r}_{0j}}) r_{0j}^n P_n(\hat{\mathbf{r}} \cdot \hat{\mathbf{r}}_{0j}). \tag{27}$$

Since we consider the surface potentials to be identical, irrespective of the number of dipoles, the coefficients of Eqs. (26) and (27) must be equal, namely

$$(\mathbf{Q} \cdot \nabla_{\mathbf{r}_0}) r_0^n P_n(\hat{\mathbf{r}} \cdot \hat{\mathbf{r}}_0) = \sum_{j=1}^{N} (\mathbf{Q}_j \cdot \nabla_{\mathbf{r}_{0j}}) r_{0j}^n P_n(\hat{\mathbf{r}} \cdot \hat{\mathbf{r}}_{0j}). \tag{28}$$

If Eq. (28) is satisfied, a distinguishing between dipoles is not possible. A thorough investigation of the latter shows that if the dipoles are parallel to each other, the uniqueness conditions (25) are fulfilled and it is impossible to decide if the measurements are induced by a single or finite number of dipoles [25].

3.1.2. Forward and inverse problem for distributed activity

If we abandon the assumption that the primary current $\mathbf{J}_\mathrm{p}$ is represented by a dipole, the additional information leading to a unique solution regarding the inverse problem is automatically lost. Notwithstanding, dropping the dipole hypothesis allows us to simulate complicated activation patterns in terms of distributed currents. The potential on the surface is computed solving Eq. (11) directly, accompanied by proper boundary conditions. For the inverse source problem, non-uniqueness remains a point of concern.

Albanese and Monk [28] illustrated that it is not possible to recreate a three-dimensional current based on EEG measurements. This result has been practically demonstrated in Ref. [27], where the authors show that the radius of a small spherical current cannot be recovered. For currents having dimensions less than three, the inverse EEG problem admits a unique solution. In previous sections, we swiftly examined currents of zero dimensionality, namely dipoles. In what follows, we will explore the forward and inverse EEG problems for one- and two-dimensional continuously distributed currents [29, 30].

We begin by assuming that the current is a small line segment of length $2L$, centred at $\mathbf{r}_0$ and oriented along an arbitrary direction $\hat{\boldsymbol{\alpha}}$. The primary current is then approximated as $\mathbf{J}_\mathrm{p} \simeq \mathbf{Q} + t\mathbf{A}$, where t is a variable taking values in the interval [-L,L], whereas $\mathbf{A}$ is the directional derivative of $\mathbf{J}_\mathrm{p}$ along $\hat{\boldsymbol{\alpha}}$. Replacing the approximation of $\mathbf{J}_\mathrm{p}$ into Eq. (11) provides the surface potential in the case of a linearly distributed current [29]. Knowledge of the surface measurements enables us to identify the position, moment, orientation and size of the current. Consequently, we have to determine 13 parameters and require a sufficient number of equations to be able to perform the identification. Operationally, the procedure in order to obtain this set of equations is in principle the same as for a single dipole. We expand the surface potential in a series of harmonic, homogeneous polynomials in Cartesian coordinates, where the coefficients of each monomial are known. Moreover, each coefficient contains a certain number of unknown parameters. Here, we must determine 13 unknowns which means we have to analytically calculate at least 13 coefficients, building the necessary system of equations. However, as the number of unknowns grows the concluding system of equations turns highly nonlinear. In the one-dimensional case, the system consists of a total of 19 equations with four constrains. It is possible to solve the latter at least semi-analytically [29].

When investigating the two-dimensional case, the mathematical complexity takes it up a notch. Assuming that the current is a small disk of radius ε, centred at $\mathbf{r}_0$ and perpendicularly oriented to the vector $\mathbf{r}_0$, the primary current can be approximated as $\mathbf{J}_\mathrm{p} \simeq \mathbf{Q} + \mathbf{r} \cdot \tilde{\mathbf{D}}$. The quantity $\tilde{\mathbf{D}}$ is called dyadic, a second-order tensor containing physical or geometrical information. The surface potential is provided replacing $\mathbf{J}_\mathrm{p}$ into Eq. (11) and solving the corresponding boundary value problem. Reaching that goal facilitates long and tedious manipulations involving integration as well [30].

The inverse problem, on the other hand, follows the guidelines outlined earlier. Due to the specific orientation of the primary current, only seven parameters have to be determined. Once more, the system of equations from which these parameters will be decided is nonlinear [30].

An independent view to the particular problem has been provided by Fokas [31]. After formulating the surface potential with explicit $\mathbf{Q}$ dependence (26), computing the surface potential for a continuously distributed current is straightforward. One has to replace $\mathbf{Q}$ by $\mathbf{J}_{\mathrm{p}}$ and integrate the resulting expression with respect to the volume of the conductor. The associated manipulations can be simplified by introducing Helmholtz decomposition for the primary current, which states that any three-dimensional smooth vector field can be resolved into the sum of an irrotational and solenoidal vector field, $\nabla\Psi(\mathbf{r}_0)$ and $\mathbf{A}(\mathbf{r}_0)$, respectively. As a result, assuming $\mathbf{J}_{\mathrm{p}} = \nabla\Psi + \nabla \times \mathbf{A}$ under the condition that $\nabla \cdot \mathbf{A} = 0$, we find

$$U_{\mathrm{surf}}(\hat{\mathbf{r}}, \mathbf{r}_0) = \oint \sum_{n=1}^{\infty} C_n^m (\Delta\Psi) r_0^n P_n(\hat{\mathbf{r}} \cdot \hat{\mathbf{r}}_0) dv(\hat{\mathbf{r}}_0). \tag{29}$$

Note that the electric potential depends only on $\Delta\Psi(\mathbf{r}_0)$. The integral in Eq. (29), obtained using Gauss theorem and integration by parts, can be computed analytically if we expand $\Psi(\mathbf{r}_0)$ in terms of spherical harmonics, namely

$$\Psi(\mathbf{r}_0) = \sum_{n=1}^{\infty} \sum_{m=-n}^{n} \psi_n^m(r_0) Y_n^m(\hat{\mathbf{r}}_0), \tag{30}$$

furnishing

$$U_{\mathrm{surf}} = \sum_{n=1}^{\infty} \sum_{m=-n}^{n} (\alpha_n \dot{\psi}_n^m - \beta_n \psi_n^m) Y_n^m(\hat{\mathbf{r}}). \tag{31}$$

It is remarkable that the above coefficients can be determined only under the assumption that $\Psi(\mathbf{r}_0)$ is a bi-harmonic function, namely a solution to the bi-harmonic operator $\Delta\Delta$. However, the precise description of the coefficients ψ_n^m remains open.

3.2. The homogeneous ellipsoidal brain

From the point of view of mathematical analysis, any three-dimensional object, such as the brain, would be best approximated with the aid of a coordinate system with three degrees of freedom, one in each direction. Fortunately, aforesaid system exists and is called the ellipsoidal coordinate system. Whereas the spherical coordinate system consists of concentric spheres centred at the origin, in the ellipsoidal coordinate system, each point is specified by the intersection of three non-degenerate second-degree surfaces, corresponding to an ellipsoid, a hyperboloid of one sheet as well as a hyperboloid of two sheets. This constitutes the ellipsoidal system significantly more complex and demanding than the spherical one (see [32] for an analytic account). For example, whereas knowledge of the brain's volume can be directly translated into the radius of the corresponding sphere, the volume of an ellipsoid is proportional to the product of three parameters a_1, a_2, a_3, the so-called semi-axes of the ellipsoid. Hence, there does not exist a unique combination of those three parameters providing the volume of the brain. In spite of the ramifications, the ellipsoidal system is an environment, which allows the installation and interpretation of analytical algorithms to a great extent as well. This stems from the fact that it is the most general system where the Laplacian operator assumes a spectral decomposition. As we dive into this particular system, we recognize that the ellipsoidal geometry is responsible for drastic variations in the behaviour of EEG when compared to the sphere.

3.2.1. *Forward and inverse problem for a single dipole and distributed activity*

In order to acquire the surface potential for an active dipole within an ellipsoidal brain, the same machinery as for the sphere is utilized. There are, however, some differences. The spherical coordinates are easily established by fixing the centre of the system and moving a distance r away. Symmetry will do the rest. In the ellipsoidal system, the procedure works a lot different. In order to solve boundary value problems in the ellipsoidal coordinate system, such as the forward EEG problem, it is essential to adopt an ellipsoid in such a way as to fit the actual boundary of the conductor under consideration, by choosing a particular value of the ellipsoidal 'radial' variable ρ. This is secured if we use the boundary of our domain to be the so-called reference ellipsoid and construct the ellipsoidal system that is based on it. Further, denote by h_1, h_2, h_3 the three semifocal distances from which the orthogonal ellipsoidal coordinate system (ρ, ν, μ) is derived. Each confocal ellipsoidal surface is defined by a constant value of the 'radial' variable $\rho \in (h_2, \infty)$, with $\rho = a_1$ indicating the reference ellipsoid.

Consider in what follows a homogeneous ellipsoidal conductor with semi-axes a_i, $i = 1,2,3$ and conductivity σ which takes the place of the reference ellipsoid for our ellipsoidal coordinate system. The surface potential is again computed by solving Eq. (11) combined with expression (14), namely

$$\Delta U = \sigma^{-1} \mathbf{Q} \cdot \nabla \delta(\mathbf{r} - \mathbf{r}_0), \; h_2 < \rho < a_1, \tag{32}$$

$$\partial_\rho U = 0, \; \rho = a_1. \tag{33}$$

Employing analytic techniques, the interested reader can find all details in Ref. [33], the solution regarding Eqs. (29) and (30) evaluated at the surface is

$$U_{\text{surf}}(\mathbf{r}; \mathbf{r}_0) = \frac{1}{\sigma} \sum_{n=1}^{\infty} \sum_{m=1}^{2n+1} \mathrm{B}_n^m(\mathbf{r}_0) \mathrm{E}_n^m(\mu) \mathrm{E}_n^m(\nu), \; \mathrm{B}_n^m(\mathbf{r}_0) = \frac{\mathbf{Q} \cdot \nabla_{\mathbf{r}_0} \mathbb{E}_n^m(\mathbf{r}_0)}{d}. \tag{34}$$

where E_n^m are the Lamé functions, $\mathbb{E}_n^m(\mathbf{r})$ symbolizes the triple product of these Lamé functions, and d some constant. Compared to the corresponding solutions for the spherical conductor, the above formula looks very similar to Eq. (17). As usual, the devil is in the detail.

Before proceeding, we mention that an elegant and straightforward expression connecting the surface potential with the moment $\mathbf{Q}$ and position $\mathbf{r}_0$ of the dipole, similar to Eq. (18), does not exist for the ellipsoidal conductor.

We turn now to the inverse problem. In principle, the procedure described for the sphere is generally applicable since it is geometry independent. Note again that every detail regarding the moment $\mathbf{Q}$ and position $\mathbf{r}_0$ of the dipole is encoded into the coefficients of expansion (31).

As a result, we expand the coefficients B_n^m for n = 1 and 2 yielding eight relations in total. Defining a new parameter g_n^m as

$$g_n^m = \mathbf{Q} \cdot \nabla_{\mathbf{r}_0} \mathbb{E}_n^m(\mathbf{r}_0) = d\ \mathrm{B}_n^m \tag{35}$$

which incorporates characteristics, such as the geometrical and physical properties of the conductor, as well as the EEG measurements, the algebraic manipulations leading to the solution are somewhat a little more painless. Solving a linear system of six equations with six unknowns, it can be shown [34] that the dipoles position $\mathbf{r}_0$ and moment $\mathbf{Q}$ depends only on $g_1^1,\ g_1^2,\ g_1^3,\ g_2^3,\ g_2^4,\ g_2^5$ as

$$x_0 = \frac{1}{2h_2h_3}\left(\frac{g_2^3}{g_1^2} + \frac{g_2^4}{g_1^3} - \frac{g_1^1 g_2^5}{g_1^2 g_1^3}\right),\ y_0 = \frac{1}{2h_1h_3}\left(\frac{g_2^3}{g_1^1} + \frac{g_2^5}{g_1^3} - \frac{g_1^2 g_2^4}{g_1^1 g_1^3}\right),\ z_0 = \frac{1}{2h_1h_2}\left(\frac{g_2^4}{g_1^1} + \frac{g_2^5}{g_1^2} - \frac{g_1^3 g_2^3}{g_1^1 g_1^2}\right) \tag{36}$$

and

$$Q_m = \frac{h_m}{h_1h_2h_3} g_1^m, \tag{37}$$

for every $m = 1,2,3$, respectively. Expressing two of the triple products $\mathbb{E}_n^m(\mathbf{r})$ in Cartesian coordinates, namely for $n = 2$ and $m = 1,2$, we obtain two constrains similar to Eq. (25), which are satisfied by the constants $g_1^1,\ g_1^2,\ g_1^3,\ g_2^3,\ g_2^4,\ g_2^5$ as

$$g_2^1 = 2(\Lambda - a_1^2)(\Lambda - a_2^2)(\Lambda - a_3^2)\left(\frac{Q_1x_0}{\Lambda - a_1^2} + \frac{Q_2y_0}{\Lambda - a_2^2} + \frac{Q_3z_0}{\Lambda - a_3^2}\right) \tag{38}$$

and

$$g_2^2 = 2(\Lambda' - a_1^2)(\Lambda' - a_2^2)(\Lambda' - a_3^2)\left(\frac{Q_1x_0}{\Lambda' - a_1^2} + \frac{Q_2y_0}{\Lambda' - a_2^2} + \frac{Q_3z_0}{\Lambda' - a_3^2}\right), \tag{39}$$

where Λ, Λ' are ellipsoidal parameters depending on the semi-axes $a_i,\ i = 1,2,3$.

In the case of a continuously distributed current, we follow the example laid out for the sphere. Replacing $\mathbf{Q}$ by the primary current $\mathbf{J}_\mathrm{p}$ in Eq. (31) and integrating over the volume of the conductor gives

$$U_{\mathrm{surf}}(\mathbf{r}; \mathbf{r}_0) = \oint \sum_{n=1}^{\infty} \sum_{m=1}^{2n+1} \mathrm{D}_n^m(\mathbf{r}_0)(\Delta\Psi)\mathrm{E}_n^m(\mu)\mathrm{E}_n^m(\nu) dv(\hat{\mathbf{r}}_0), \tag{40}$$

where Helmholtz's decomposition has been used. We compute the above integral analytically by expanding $\Psi(\mathbf{r}_0)$ in terms of the product of Lamè functions $\mathrm{E}_n^m(\mu)\mathrm{E}_n^m(\nu)$, namely

$$\Psi(\mathbf{r}) = \sum_{n=1}^{\infty} \sum_{m=1}^{2n+1} \psi_n^m(\rho)\mathrm{E}_n^m(\mu)\mathrm{E}_n^m(\nu). \tag{41}$$

After a series of cumbersome algebra, it can be shown that [31]

http://dx.doi.org/10.5772/68021

$$U_{\text{surf}} = \sum_{n=1}^{\infty}\sum_{m=1}^{2n+1} \mathrm{D}_n^m(\mathbf{r}_0) I_n^m \, \mathrm{E}_n^m(\mu)\mathrm{E}_n^m(\nu), \tag{42}$$

where

$$I_n^m = \int_{h_2}^{a_1} \psi_n^m(\rho)\frac{\mathrm{d}}{\mathrm{d}\rho}\left(\sqrt{\rho^2 - h_3^2}\sqrt{\rho^2 - h_2^2}\frac{\mathrm{d}}{\mathrm{d}\rho}E_n^m(\rho)\right)\mathrm{d}\rho$$

$$+\sqrt{{a_1}^2 - h_3^2}\sqrt{{a_1}^2 - h_2^2}\left(E_n^m(a_1)\dot{\psi}_n^m - \dot{E}_n^m(a_1)\psi_n^m\right). \tag{43}$$

As for the spherical conductor, additional information is required in order to evaluate the coefficients ψ_n^m. A possible approach to uniquely determine the coefficients ψ_n^m is to consider some form of minimizer. For example, a widely used concept in medical imaging is the minimum principle, the assumption that the current should have minimum 'strength', mathematically expressed as the minimization of the L_2 norm of $\mathbf{J}_\mathrm{p}$.

4. Discussion

We presented a brief but concise introduction of the mathematical terminology associated with the modality of EEG. Having in mind medical and health professionals, we start from the very beginning, presenting step by step the physics and mathematical formulation behind EEG in a simple manner, keeping the mathematical notation to a minimum. The tools and techniques needed in order to derive at the presented results are intentionally not incorporated for two reasons. First of all, the procedure deriving at these formulas is not an easy task in general. Secondly, the main focus of this work is to display the beauty of the final expressions for every problem, showing how every single piece of information is encoded within these formulas and by what means the extraction of conclusions is accomplished.

Our introduction starts with the most elementary model possible, which, at the same time, is also the most straightforward and understandable of models. Representing the brain as a homogeneous sphere is an unrealistic assumption but serves an important task. At present, it is the only geometry for which the electromagnetic fields generated by a dipole source are exactly known in closed form. Further, it is needed in order to be able to draw conclusions when we move to build models of higher complexity. It therefore reveals disparities between forthcoming models, but more substantially, it allows us to test the reliability of the introduced algorithms in a straightforward and timely matter. Moreover, an analytic benchmark problem is provided, which can be used to test existing and new formulations.

For example, based on the homogeneous spherical model it is shown in Ref. [35] to what degree deformations present at the conductor's surface affect EEG measurements. Although the EEG data are evaluated in a view to a deformed conductor, the calculations are accomplished based on the spherical geometry, furnishing a fast analytic algorithm prone to almost minimum error. Another characteristic example of rigorous mathematical analysis is the quantitative description of the non-uniqueness for the EEG inverse problem, presented in Ref. [36]. Therein, splitting the current into components, the authors prove that none of those components contributes to both the electric and scalar magnetic potential; in other words, recordings of EEG and MEG do not contain overlapping information about the current. However, aforementioned property holds no longer true if the spherical conductor is disregarded.

Analogous conclusions are valid for the ellipsoidal geometry as well. For example, the authors consider in Ref. [37] the frequent case when clinical data of unknown origin are implemented in computational simulations. We mentioned earlier that there exist a plethora of combinations of the product a_1, a_2, a_3 furnishing the same value for the volume of the brain. If we look at the instance where EEG measurements originate from a brain with fixed values a_1, a_2, a_3 but are interpreted in the sequel by different values, what would be the error? Well, it turns out that the error can reach as high as 20%, depending on the position and strength of the primary current.

The error analysis presented in Ref. [37] can be considered as rather straightforward, since both ellipsoids under consideration were confocal, that is, members of the same ellipsoidal system enjoying the same foci. In plain words, no member of a confocal family touches another. Consequently, there exists a single curve, which cuts both ellipsoids normally, and the corresponding intersecting points consist of the most proximate pair between the two ellipsoids in each direction. These two points are employed in the analysis calculating the aforementioned error. But what would happen if the two ellipsoids would not be considered to be confocal? This interesting case is also more realistic.

In order to provide an answer to the latter, a sophisticated correspondence is needed connecting two points on the surface of two ellipsoids which are now non-confocal. This means that there probably exists a point shared by both ellipsoids. In Ref. [38], the authors investigated the effect implied by a deviation of the eccentricities of the ellipsoidal model on the electric potentials registered as the EEG data. In this case, the error reaches high values up to almost 100%.

Turning our attention from the geometrical deformation of the conductor model, to the physical assumption of homogeneity, we acknowledge the significance of the non-homogeneity imposed by the layers of different conductivity that cover the host tissue of the EEG source. The conductive elements that constitute the scalp, the scull and the meninges, which interfere between the EEG measurements and the cerebrum, are affected by the electromagnetic field produced by activation of the source. Hence, they induce a volume current that perturbs the total electric potential registered on the EEG receptors on the scalp. The effect of this physical perturbation of the potential has been studied by incorporating a layered conductivity profile in all the models discussed in the present review, by characterizing each layer by a distinct but constant conductivity value.

Indicatively, we infer that switching to a layered ellipsoidal model of the head-brain system, the functional form of the electric potential, is basically unaltered. One of the authors has showed [39] that the conductivity profile of the layered structure enters the potential formula by normalizing each term by a constant which incorporates the conductivity jumps across the interfaces and the geometrical characteristics of the layers. Analogous results show a similar effect on the inhomogeneous conductor [31]. Hence, the formula of the electric potential that will serve as the stepping stone for the inverse calculations is the one that corresponds to the most realistic inhomogeneous models that acknowledge the layered conductivity profile of the head-brain system. A promising challenge for future investigations refers to incorporating the anisotropic conductivity profile, where the conductivity varies with the direction into each separate compartment, modelling, for example, the different conductivity of the white matter of the brain than that of grey matter.

Using tensor conductivity for modelling the brain anisotropy is one of the many analytical mathematical challenges of this fascinating area of functional brain imaging while creative mathematical modelling has a lot to contribute in the close examination of EEG theory and applications.

Author details

Michael Doschoris[1]* and Foteini Kariotou[2]

*Address all correspondence to: mdoscho@chemeng.upatras.gr

1 Department of Chemical Engineering, University of Patras, University Campus, Patras, Hellas

2 School of Science & Technology, Hellenic Open University, Patras, Hellas

References

[1] M.F. Glasser, T.S. Coalson, E.C. Robinson, C.D. Hacker, J. Harwell, E. Yacoub, K. Ugurbil, J. Andersson, C.F. Beckmann, M. Jenkinson, S.M. Smith and D.C. Van Essen. A multi-modal parcellation of human cerebral cortex. Nature. 2016;**536**:171–178. DOI: 10.1038/nature18933

[2] T. Vanderah and D. Gould. Nolte's The Human Brain: An Introduction to its Functional Anatomy. 7th ed. Elsevier; 2015. 720 p. ISBN-13: 978-1455728596

[3] E.R. Kandel, J.H. Schwartz, T.M. Jessell. Principles of Neural Science. 4th ed. McGraw-Hill; 2000. 1414 p. ISBN-13: 978-0838577011

[4] B.J. Larson. A review of the history of electrophysiology and electroencephalography: Part I. American Journal of EEG Technology. 1963;**3**(1):13–16. DOI: 10.1080/00029238.1963.11080601

[5] B.J. Larson. A review of the history of electrophysiology and electroencephalography: Part II. American Journal of EEG Technology. 1963;**3**(3):61–64. DOI: 10.1080/00029238.1963.11080609

[6] M. Scherg. Fundamentals of dipole source potentials. In: F. Grandori, M. Hoke, and G.I. Romani, editors. Auditory Evoked Magnetic Fields and Electric Potentials. Karger; 1990. p. 40–69. ISBN-13: 978-3805550017

[7] P.L. Nunez and R. Srinivasan. Electric Fields of the Brain: The Neurophysics of EEG. 2nd ed. Oxford University Press; 2006. 640 p. ISBN-13: 978-0195050387

[8] H. Hallez, B. Vanrumste, R. Grech, J. Muscat, W. De Clercq, A. Vergult, Y. D'Asseler, K.P. Camilleri, S.G. Fabri, S. Van Huffel, and I. Lemahieu. Review on solving the forward problem in EEG source analysis. Journal of NeuroEngineering and Rehabilitation. 2007;**4**(1): 46. DOI: 10.1186/1743-0003-4-46

[9] R. Grech, T. Cassar, J. Muscat, K.P. Camilleri, S.G. Fabri, M. Zervakis, P. Xanthopoulos, V. Sakkalis, and B. Vanrumste. Review on solving the inverse problem in EEG source analysis. Journal of NeuroEngineering and Rehabilitation. 2008;**5**(1):25. DOI: 10.1186/1743-0003-5-25

[10] S. Baillet, J. Mosher, and R. Leahy. Electromagnetic brain mapping. IEEE Signal Processing Magazine. 2001;**18**(6):14–30. DOI: 10.1109/79.962275

[11] F. Vatta, F. Meneghini, F. Esposito, S. Mininel, and F. Di Salle. Realistic and spherical head modeling for EEG forward problem solution: a comparative cortex-based analysis. Computational Intelligence and Neuroscience. 2010;**2010**:972060, 11 pages. DOI: 10.1155/2010/972060

[12] F.N. Wilson and R.H. Bayley. The electric field of an eccentric dipole in a homogeneous spherical conducting medium. Circulation. 1950;**1**(1):84–92.

[13] B. Vanrumste, G. Van Hoey, R. Van de Walle, M.R.P. D'Havé, I.A. Lemahieu, and P.A.J.M. Boon. Comparison of performance of spherical and realistic head models in dipole localization from noisy EEG. Medical Engineering & Physics. 2002;**24**(6):403–418. DOI: 10.1016/S1350-4533(02)00036-X

[14] Mosher, R. Leahy, P. Lewis. EEG and MEG: forward solutions for inverse methods. IEEE Transactions on Biomedical Engineering. 1999;**46**(3):245–259.

[15] Z. Akalin Acar, S. Makeig. Effects of forward model errors on EEG source localization. Brain Topography. 2013;**26**(3):378–396. DOI: 10.1007/s10548-012-0274-6.

[16] J. Vorwerk, J.H. Cho, S. Rampp, H. Hamer, T.R. Knösche, C.H. Wolters. A guideline for head volume conductor modeling in EEG and MEG. NeuroImage. 2014;**100**:590–607. DOI: 10.1016/j.neuroimage.2014.06.040

[17] Y. Huang, L.C. Parra and S. Haufe. The New York Head—A precise standardized volume conductor model for EEG source localization and tES targeting. NeuroImage. 2016;**140**:150–162. DOI: 10.1016/j.neuroimage.2015.12.019

[18] L.D. Landau, L.P. Pitaevskii and E.M. Lifshitz. Electrodynamics of Continuous Media. 2nd ed. Butterworth-Heinemann; 1984. 460 p. ISBN-13: 978-0750626347

[19] P. Plonsey and D.B. Heppner. Considerations of quasi-stationarity in electrophysiological systems. Bulletin of Mathematical Biophysics. 1967;**29**(4):657–664. DOI: 10.1007/BF02476917

[20] O. Darrigol. Electrodynamics from Ampère to Einstein. Oxford University Press; 2003. 552 p. ISBN-13: 978-0198505938

[21] J.I. Richards and H.K. Youn. The Theory of Distributions: A Nontechnical Introduction. Cambridge University Press; 1995. 157 p. ISBN-13: 9780521371490

[22] G. Dassios. Electric and magnetic activity of the brain in spherical and ellipsoidal geometry. In: H. Ammari, editor. Mathematical Modeling in Biomedical Imaging I. Electrical and Ultrasound Tomographies, Anomaly Detection, and Brain Imaging. Springer; 2009. p. 133–202. DOI: 10.1007/978-3-642-03444-2_4

[23] F.W.J. Olver, D.W. Lozier, R.F. Boisvert, C.W. Clark, editors. NIST Handbook of Mathematical Functions. Cambridge University Press; 2010. 968 p. ISBN-13: 978-0521140638

[24] D. Yao. Electric potential produced by a dipole in a homogeneous conducting sphere. IEEE Transactions on Biomedical Engineering. 2000;**47**(7):964–966. DOI: 10.1109/10.846691

[25] G. Dassios, M. Doschoris and K. Satrazemi. Localizing brain activity from multiple distinct sources via EEG. Journal of Applied Mathematics. 2014;**2014**:232747. DOI: 10.1155/2014/232747

[26] A. Ben-Israel, T.N.E. Greville. Generalized Inverses: Theory and Applications. 2nd ed. Springer; 2003. 420 p. ISBN-13: 978-0387002934

[27] G. Dassios and A.S. Fokas. Electro-magneto-encephalography for a three-shell model: dipoles and beyond for the spherical geometry. Inverse Problems. 2009;**25**(3):035001. DOI: 10.1088/0266-5611/25/3/035001

[28] R. Albanese and P.B. Monk. The inverse source problem for Maxwell's equations. Inverse Problems. 2006;**22**(3):1023–1035. DOI: 10.1088/0266-5611/22/3/018

[29] G. Dassios, G. Fragoyiannis, and K. Satrazemi. On the inverse EEG problem for a 1D current distribution. Journal of Applied Mathematics. 2014;**2014**: ID 715785, 11 pages. DOI: 10.1155/2014/715785

[30] G. Dassios and K. Satrazemi. Inversion of electroencephalography data for a 2-D current distribution. Mathematical Methods in the Applied Sciences. 2014;**38**(6):1098–1105. DOI: 10.1002/mma.3132

[31] A.S. Fokas. Electro-magneto-encephalography for a three-shell model: distributed current in arbitrary, spherical and ellipsoidal geometries. Journal of The Royal Society Interface. 2009;**6**(34):479–488. DOI: 10.1098/rsif.2008.0309

[32] G. Dassios. Ellipsoidal Harmonics. Theory and Applications. 1st ed. Cambridge University Press; 2012. 474 p. ISBN-13: 9780521113090

[33] F. Kariotou. Electroencephalography in ellipsoidal geometry. Journal of Mathematical Analysis and Applications. 2004;**290**(1):324–342. DOI: 10.1016/j.jmaa.2003.09.066

[34] G. Dassios and A.S. Fokas. Electro-magneto-encephalography for the three-shell model: a single dipole in ellipsoidal geometry. Mathematical Methods in the Applied Sciences. 2012;**35**(12):1415–1422. DOI: 10.1002/mma.2508

[35] M. Doschoris, G. Dassios, and G. Fragoyiannis. Sensitivity analysis of the forward electroencephalographic problem depending on head shape variations. Mathematical Problems in Engineering. 2015;**2015**:612528, 14 pages. DOI: 10.1155/2015/612528

[36] G. Dassios and D. Hadjiloizi. On the non-uniqueness of the inverse problem associated with electroencephalography. Inverse Problems. 2009;**25**(11):115012, 18 pp. DOI: 10.1088/0266-5611/25/11/115012

[37] M. Doschoris and F. Kariotou. Quantifying errors during the source localization process in Electroencephalography. Part 1: Confocal systems. IMA Journal of Applied Mathematics. In revision.

[38] M. Doschoris and F. Kariotou. Quantifying errors during the source localization process in Electroencephalography. Part 2: Non-confocal systems. Under review.

[39] S.N. Giapalaki and F. Kariotou. The complete ellipsoidal shell model in EEG imaging. Abstract and Applied Analysis. 2006;**2006**:57429, 18 pages. DOI: 10.1155/AAA/2006/57429

Chapter 5

Filter Band Multicarrier Based Transmission Technology for Clinical EEG Signals

Chin-Feng Lin, Wei-Syuan Chao and Jun-Da Chen

Additional information is available at the end of the chapter

http://dx.doi.org/10.5772/intechopen.68838

Abstract

A transmission scheme is proposed based on filter band multicarrier (FBMC) transmission technology for clinical electroencephalogram (EEG) signals. The proposed scheme integrates binary phase shift keying (BPSK) and offset quadrature amplitude modulation (OQAM), an FBMC transmission mechanism, and low-density parity-check code (LDPC) error protection in an FBMC-based EEG mobile communication system. The proposed EEG mobile communication system employs high-speed transmission, with schemes providing significant error protection for mobile communication of clinical EEG signals requiring a stringent bit-error rate (BER). The performances of BERs and mean square errors (MSEs) of the proposed EEG mobile communication system were explored. Simulation results show that the proposed scheme is a superior transmission platform as compared to existing schemes for clinical EEG signals.

Keywords: EEG, FBMC, OQAM, LDPC

1. Introduction

Noninvasive monitoring of brain activity in daily life is an important research topic in the field of healthcare that aims at a comfortable lifestyle. Mihajlovic et al. [1] demonstrated various aspects of wireless and intelligent wearable lifestyle electroencephalogram (EEG) solutions, and the technology behind the development of convenient, intelligent, and wearable, wireless EEG devices was explored. In addition, in their study, personality traits, sensory input, neuronal activity, conductive tissues, electrode-tissue interface, miniaturized and ergonomic EEG headsets, wireless and wearable EEG system designs, brain activity analysis, and output interfaces were discussed. Lin et al [2] proposed a Bluetooth-based real-time brain-computer interface (BCI) system that can be used to detect drowsiness while driving. This system had integrated wireless physiological signal-acquisition and embedded signal-processing modules, and it featured real-time wireless drowsiness detection, long-term daily life EEG monitoring, high computation capacity, and low power consumption. The design and implementation of a Bluetooth-based wearable brain monitoring system was investigated by Sawan et al. [3]. A wireless data recording system was utilized for noninvasive and long-term monitoring of near-infrared spectrometry (NIRS) EEG signals. In addition, the wireless data recording system was applied to the field of invasive cerebral EEG detection. This system has a graphical user interface that is user-friendly and can be used to extract brain activity during dynamic tasks. The advantages of the designed system are portability, wireless connectivity, high throughput, reliable communication, and low-power consumption. Liao et al. [4] proposed a design method for the 16-channel EEG measurement system utilizing dry spring-loaded sensors, a Bluetooth-based acquisition system, and a size-adjustable wearable soft cap. Vos et al. [5] demonstrated an efficient low-cost mobile EEG system that utilizes a P300-based speller for wireless BCI.

The development of high-speed, and reliable EEG transmission schemes are interesting research topics. Channel coding is a solution that can be used to achieve a lower error probability for EEG communication. The fundamental design parameters of channel coding are error probability, complexity, and decoding time. Low-density parity-check (LDPC) code is a channel coding technology that was proposed by Gallager [6]. Limpaphayom et al. [7] proposed a power and bandwidth-efficient communication system that utilizes irregular LDPC component codes of block length 100,000, multilevel coding, multistage decoding, 64-quadrature amplitude modulation, and trellis-based signal shaping schemes. The proposed system achieved a bit-error rate (BER) of 10^{-5} at an Eb/No of 6.55 dB. Franceschini et al. [8] described the concept ofLDPC codes. The regular (v, c) LDPC codes are linear block codes with a sparse parity-check matrix H, In H, the number of nonzero elements in the columns is v, while the number of nonzero elements in the rows is c. In addition, the code rate is defined as 1 − v/c. In an (N,K) LDPC code, the block length is N and the information length is K. Ohtsuki [9] applied LDPC codes to various transmission systems with excellent performance, and illustrated some of the LDPC code designs. LDPC codes are included in the second-generation specification for satellite broadband applications, and in IEEE 802.16e.

Filter bank-based multicarrier modulations (FBMC) is an interesting research topic, and it is being considered as a potential candidate for the fifth generation (5G) mobile systems. FMBC is a modified version of orthogonal frequency division multiplexing (OFDM), and a tutorial review of FBMC modulations was discussed by Boroujeny et al. [10]. Compared to cyclic prefix (CP)-based OFDM modulation, FBMC offers better spectral efficiency in multipath channels. Bouhadda et al. investigated the BER performance of nonlinear distortion in high-power amplifiers for FBMC using offset quadrature amplitude modulation (OQAM) [11]. Caus et al. [12] studied the effects of multi-tap filtering on FBMC/OQAM systems to combat intersymbol and inter-carrier interferences due to multipath fading. Caus et al. [12] proposed a low-complexity transmission power estimation method, and their simulation results show that the proposed transmission power estimation method is excellent. Further, Bellanger et al. [13] proposed an FBMC-based physical layer solution for 5G mobile systems.

http://dx.doi.org/10.5772/intechopen.68838

In previous studies, a survey study of mobile telemedicine [14], mobile telemedicine using an advanced wireless multimedia communication application [15], an 802.11n wireless telemedicine application [16], a direct sequence ultra-wideband (DS-UWB) wireless telemedicine application [17], a multi-code code division multiple access (CDMA) mobile medicine system [18], a Ka band OFDM-based multi-satellite mobile telemedicine system [19], a Ka band wideband CDMA mobile telemedicine system [20], and a mobile cloud-based blood pressure healthcare system [21] were investigated. In this chapter, an advanced FBMC-based EEG transmission scheme is proposed. The design concept of the proposed advanced wireless EEG transmission system includes FBMC, LDPC, BPSK or OQAM, and a power assignment mechanism. Low power high-speed wireless EEG transmission was achieved.

2. System model

The proposed FBMC-based EEG transmission scheme is shown in **Figure 1**. The LDPC channel coding, FBMC transmission method, a BPSK or OQAM adaptive modulation, and a power assignment mechanism were used in a new design strategy to achieve low power, high speed, and high quality of service transmission capabilities for wireless EEG systems. We performed a simulation using the irregular LDPC codes set with a block length of 2000 and a rate of 1/2. The digital bit streams of the EEG signal were inputted to the irregular LDPC encoder, and the LDPC bit streams were outputted. The OQAM modulation scheme is used at the lower fading channel in order to achieve high-speed wireless EEG transmission. A BPSK is used at the highest fading channel, to achieve robust EEG communication. The LDPC bit streams were inputted to an OQAM-based or BPSK-based adaptive modulator, and the LDPC bit streams with adaptive modulation were outputted. The LDPC bit streams with adaptive modulation were inputted to the serial to parallel converter, and the 64 parallel LDPC adaptive modulation EEG bit streams were outputted. The 64 parallel LDPC adaptive modulation EEG bit streams were inputted to the digital filter banks (DFBs). The coefficients of the DFBs were determined to be as follows [13].

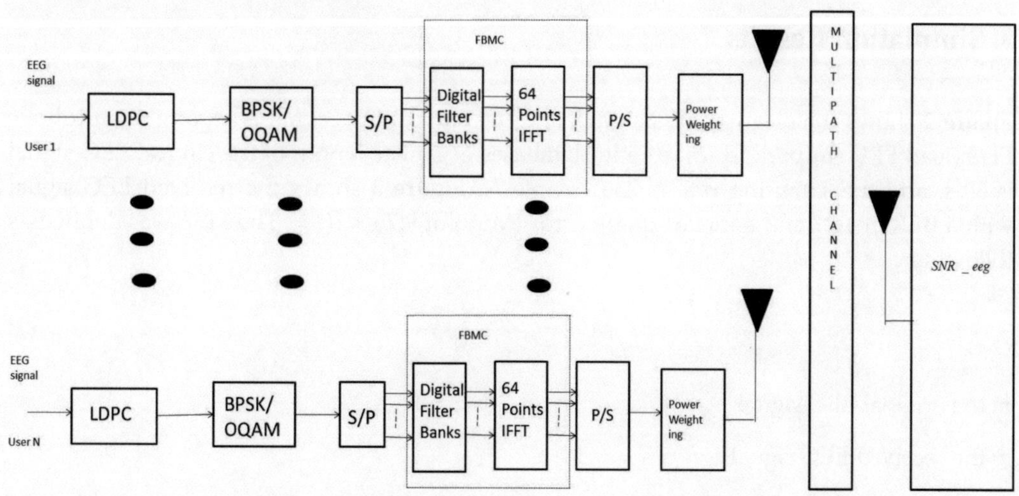

Figure 1. The proposed FBMC-based EEG transmission scheme.

$$H_0 = 1, H_1 = 0.971960, H_2 = \sqrt{2}/2, H_3 = 0.235147 \tag{1}$$

Further, the 64 DFB output bit streams were inputted to a 64-point inverse fast Fourier transform (IFFT), and the 64 parallel LDPC adaptive modulation FBMC EEG bit streams were outputted. The 64 parallel LDPC adaptive modulation FBMC EEG bit streams were inputted to a parallel to serial converter, and the serial LDPC adaptive modulation FBMC EEG bit streams were output. The proposed power assignment algorithm is summarized below:

Step 1: On the basis of the output information obtained from the object-component Petri Nets (OCPN) model, determine throughputs for EEG packets transmission.

Step 2: Select the appropriate modulation mode to satisfy the requirements for transmission over a wireless EEG transmission system.

Step 3: Assign the initial value of power weighting as 15/30 for EEG packets.

Step 4: Measure the received signal to noise ratio (SNR) for EEG packets.

Step 5: If the measured SNR of the received signal exceeds the threshold SNR at which the required BER for EEG packets is achieved, then update the power weighting as power weighting = power weighting − 1/30; If power weighting ≥1/30, go to Step 4; otherwise, go to Step 7.

Step 6: If the measured SNR of the received signal is less than the threshold SNR at which the required BER for EEG packets is achieved, then update the power weighting as power weighting = power weighting + 1/30; If power weighting ≤1, go to Step 4; otherwise, go to Step 8.

Step 7: Change the modulation mode. If the modulation mode is not OQAM, go to Step 4.

Step 8: Increase the modulation mode. If the modulation mode is not BPSK, go to Step 4.

The carrier sense multiple access with collision avoidance (CSMA/CA) technology were used for multiuser wireless EEG transmission system.

3. Simulation Results

Figure 2 shows the original EEG signal, and the clinical test signal is an original alcoholic EEG (lead FP1) (http://kdd.ics.uci.edu/databases/eeg). The length of the clinical EEG signal is 50 s, and the sampling rate is 256 samples/s. **Figure 3** shows the received EEG signal with a BER of 10^{-7} and a mean square error (MSE) of 9.76×10^{-10}. The MSE was defined as following:

$$MSE = \frac{\sum_{i=1}^{N} (x_i - y_i)^2}{N} \tag{2}$$

x_i: the original EEG signal.

y_i: the received EEG signal.

N: the length of original EEG signal.

http://dx.doi.org/10.5772/intechopen.68838

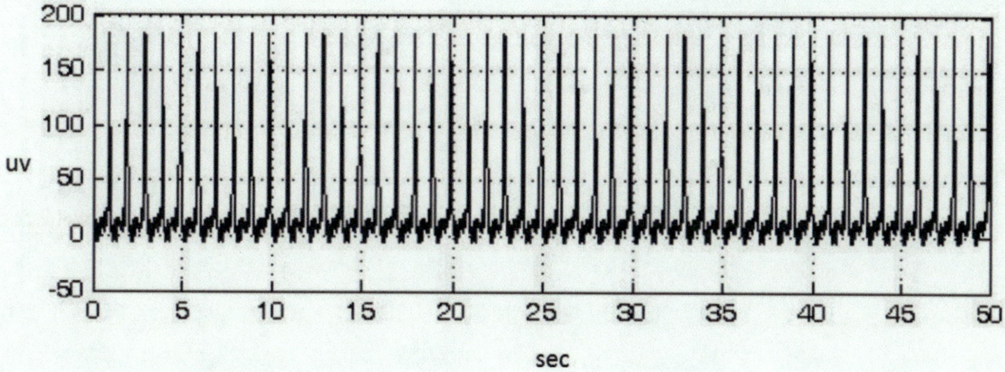

Figure 2. The original EEG signal.

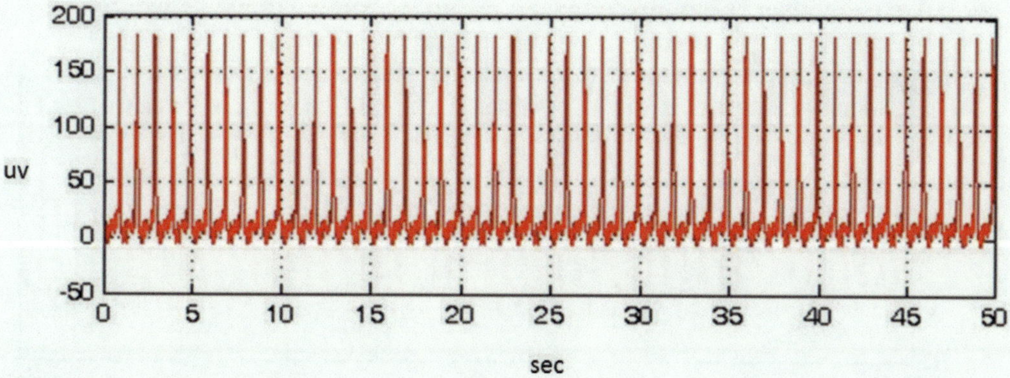

Figure 3. The received EEG signal with a BER of 10^{-7} and an MSE of 9.76×10^{-10}.

The block length of 2000 and rate of 1/2 irregular LDPC encoder, BPSK, and 64-points IFFT were used in the simulation, and the coefficients of the tap-delay line multipath channel model were [1, 0.5, 0.25, 0.125]. The assumed coefficients of multipath channel estimation (ACMCE) were [0.99, 0.495, 0.2475, 0.12375]. **Figure 4** shows the received EEG signal with a BER of 2.5×10^{-6} and an MSE of 5.24×10^{-5}. **Figure 5** shows the received EEG signal with a BER of 2.5×10^{-5} and an MSE of 1.32×10^{-2}. There was no difference on the human vision. **Figure 6** shows the received EEG signal with a BER of 1.7×10^{-3} and an MSE of 0.21. **Figure 7** shows the received EEG signal with a BER of 8.8×10^{-3} and an MSE of 1044.10. There were obviously differences on the human vision.

Our BER performance results for the proposed wireless EEG transmission system are shown in **Figure 8**. The six signs "△," "□," "o," "☆," "*," and "x" denote the ACMCE with no error using BPSK, ACMCE with no error using OQAM, ACMCE with 1% error using BPSK, ACMCE with 1% error using OQAM, ACMCE with 10% error using BPSK, and ACMCE with 10% error using OQAM, respectively. As shown in the figure, the BER of BPSK was smaller than that of OQAM. Furthermore, the BER of the ACMCE with 10% error was larger than that of the ACMCE with no error. The MSE performance results for the proposed wireless EEG

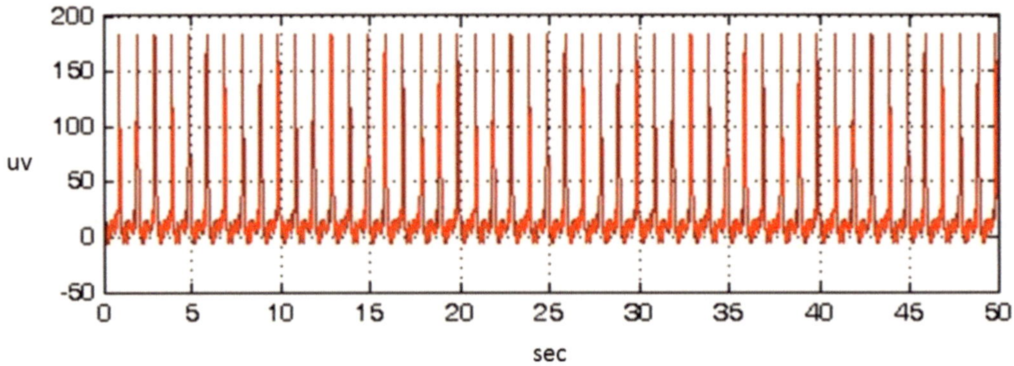

Figure 4. The received EEG signal with a BER of 2.5×10^{-6} and an MSE of 5.24×10^{-5}.

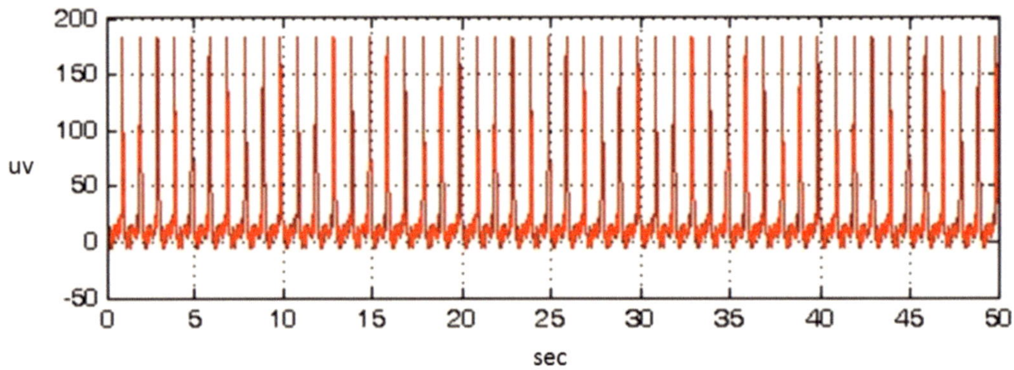

Figure 5. The received EEG signal with a BER of 2.5×10^{-5} and an MSE of 1.32×10^{-2}.

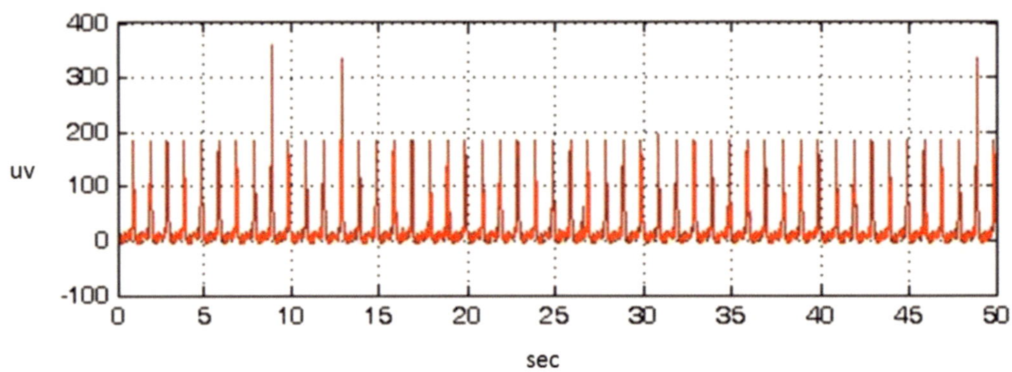

Figure 6. The received EEG signal with a BER of 1.7×10^{-3} and an MSE of 0.21.

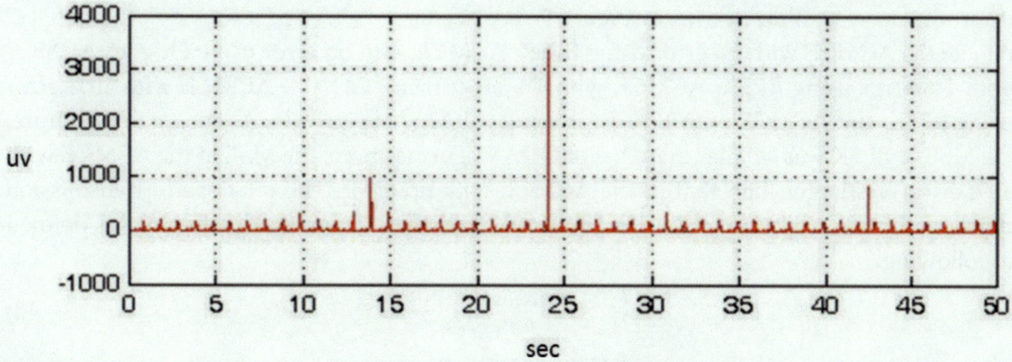

Figure 7. The received EEG signal with a BER of 8.8×10^{-3} and an MSE of 1044.10.

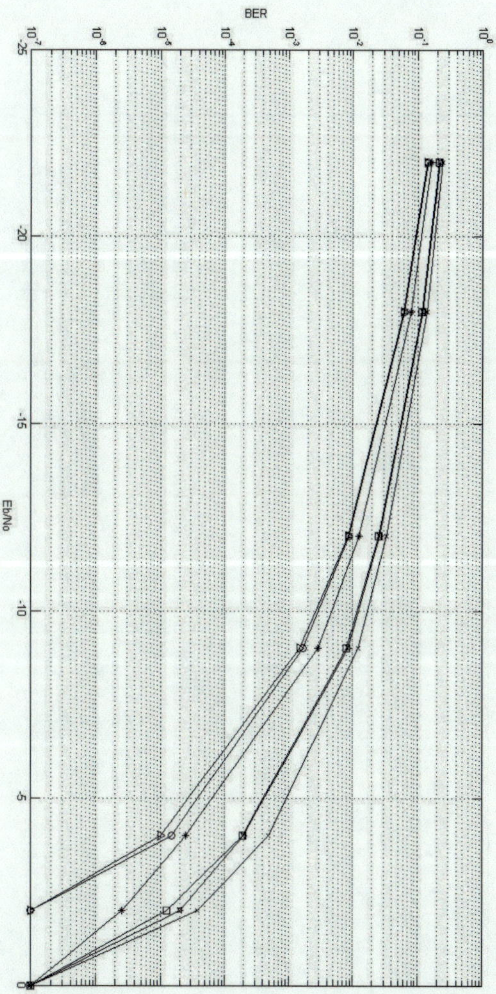

Figure 8. The BER performance results for the proposed wireless EEG transmission system.

transmission system are shown in **Figure 9**. The six signs "△," "□," "o," "☆," "*," and "x" denote the ACMCE with no error using BPSK, ACMCE with no error using OQAM, ACMCE with 1% error using BPSK, ACMCE with 1% error using OQAM, ACMCE with 10% error using BPSK, and ACMCE with 10% error using OQAM, respectively. As shown in the figure, the MSE of BPSK was smaller than that of OQAM. Furthermore, the MSE of the ACMCE with 10% error was larger than that of the ACMCE with no error. The relations of transmission power and noise power are shown in **Figure 10**. The signal to noise ratio, Eb/No, was defined as following:

$$\frac{E_b}{N_o} = \frac{transmission\ \ power}{noise\ \ power} \tag{3}$$

$$E_b\!:\!transmission\ \ power$$

$$N_o\!:\!noise\ \ power$$

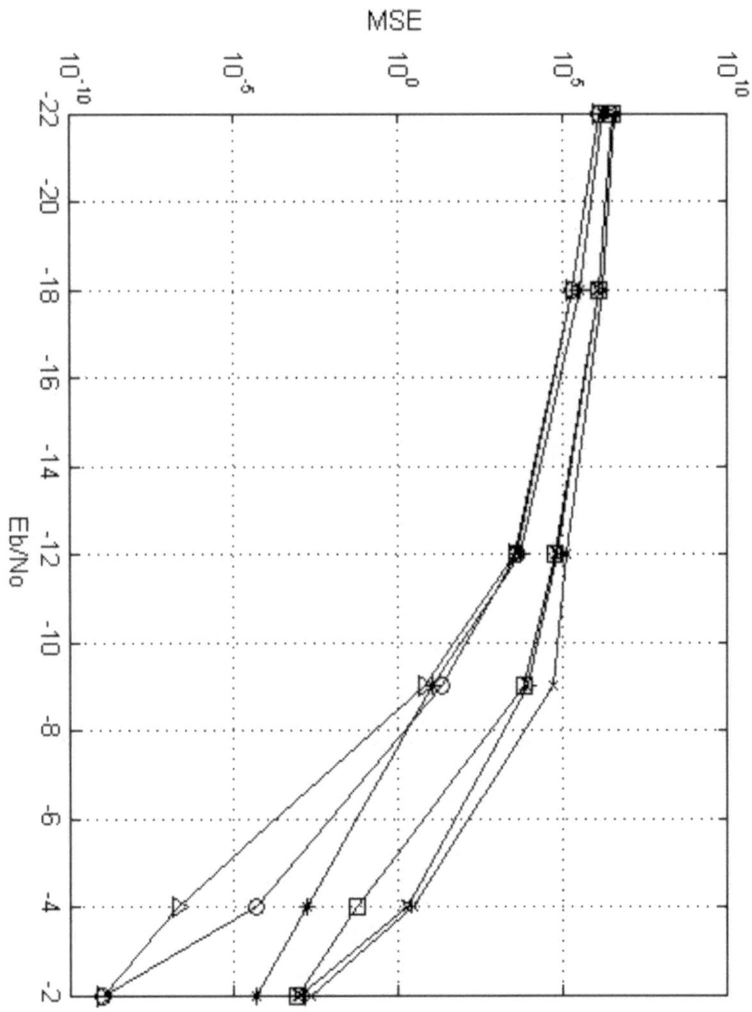

Figure 9. The MSE performance results for the proposed wireless EEG transmission system.

The transmission BER was 10^{-7}. The four signs "o," "x," "△," and "☆," denote the ACMCE with no error using BPSK, ACMCE with no error using OQAM, ACMCE with 15% error using BPSK, and ACMCE with 15% error using OQAM, respectively. When noise power was fixed, the transmission power of ACMCE with 15% error using OQAM was the highest, the transmission power of ACMCE with 15% error using BPSK was the second highest, the transmission power of ACMCE with no error using OQAM was the third highest, and the transmission power of ACMCE with no error using BPSK was the lowest. The relation of ratios of the saving transmission power and noise power were shown in **Figure 11**. The saving in transmission power was defined as following:

$$power \quad saving = (1 - E_b) * 100\% \tag{4}$$

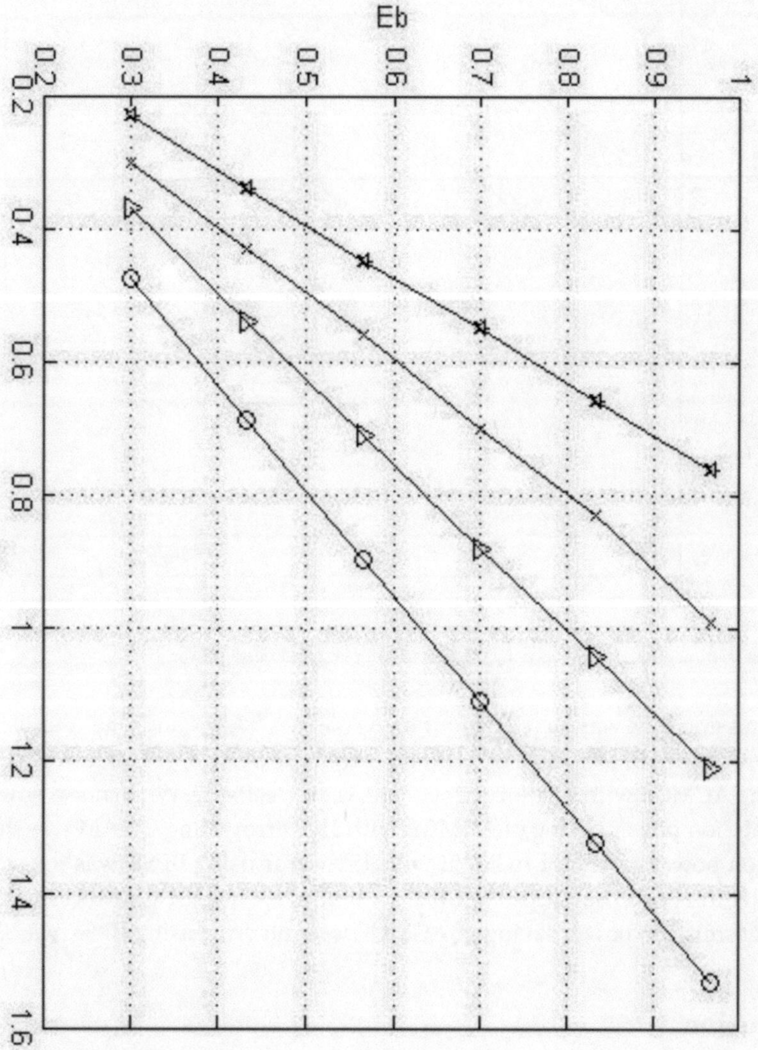

Figure 10. The relations of transmission power and noise power.

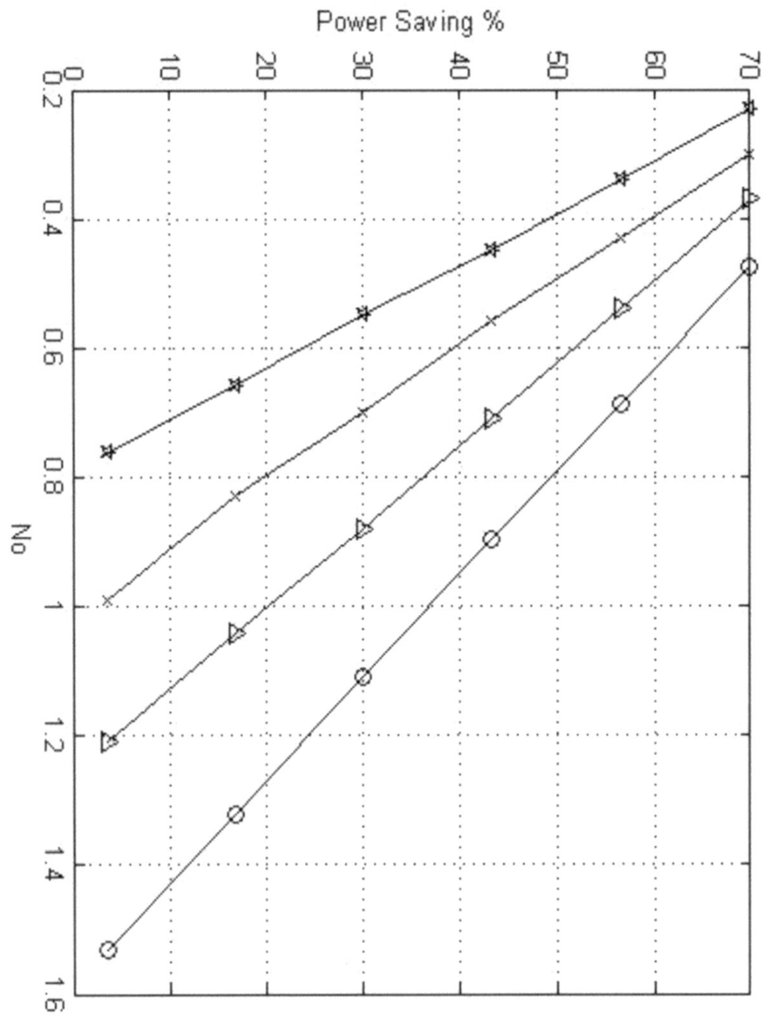

Figure 11. The relation of ratios of the saving transmission power and noise power.

The transmission BER was 10^{-7}. The four signs "o," "x," "△," and "☆," denote the ACMCE with no error using BPSK, ACMCE with no error and OQAM, ACMCE with 15% error and BPSK, using ACMCE with 15% error and OQAM, respectively. When noise power was fixed, the transmission power saving of ACMCE with 15% error using OQAM was the lowest, the transmission power saving of ACMCE with 15% error using BPSK was the second lowest, the transmission power saving of ACMCE with no error using OQAM was the third lowest, and the transmission power saving of ACMCE with no error using BPSK was the highest.

4. Conclusion

In this chapter, a new FBMC-based wireless EEG transmission technology was investigated. Advanced LDPC, OQAM and BPSK adaptive modulation, FBMC, wireless communication methods and a power assignment mechanism were integrated. Simulation results were obtained for a variety of received EEG signals, with a range of BER results. The BER performances of the proposed wireless EEG transmission architectures and the MSE performances of the proposed wireless EEG transmission architectures were demonstrated. The beneficial effects of the ACMCE errors were indicated. When the transmission BERs were 10^{-7}, the relationship between transmission power and noise power, and the relationship between the saving transmission power ratios and noise power were discussed. The simulation results confirm that the proposed transmission system is an excellent wireless EEG platform that provides the high speed and low power consumption benefits sought.

Acknowledgements

The authors acknowledge the support of the teching alliance of the Ministry of Education for Medical Electronics in Taiwan, MOST 105-2221-E-019-021, and the valuable comments of the reviewers.

Author details

Chin-Feng Lin*, Wei-Syuan Chao and Jun-Da Chen

*Address all correspondence to: lcf1024@mail.ntou.edu.tw

Department of Electrical Engineering, National Taiwan Ocean University, Taiwan

References

[1] Mihajlovic V, Grundlehner B, Vullers R, et al. Wearable, wireless EEG solutions in daily life applications: What are we missing? IEEE Journal of Biomedical and Health Informatics. 2015;**19**(1):6-21. DOI: 10.1109/JBHI.2014.2328317

[2] Lin CT, Chang CJ, Lin BS, et al. A real-time wireless brain–computer interface system for drowsiness detection. IEEE Transactions on Biomedical Circuits and Systems. 2010;**4**(4):214-222. DOI: 10.1109/TBCAS.2010.2046415

[3] Sawan M, Salam MT, Lan IL, et al. Wireless recording systems: From noninvasive EEG-NIRS to invasive EEG devices. IEEE Transactions on Biomedical Circuits and Systems. 2013;**7**(2):186-195. DOI: 10.1109/TBCAS.2013.2255595

[4] Liao LD, Wu SL, Liou CH, et al. A novel 16-channel wireless system for electroencephalography measurements with dry spring-loaded sensors. IEEE Transactions on Instrumentation and Measurement. 2014;**63**(6):1545-1555. DOI: 10.1109/TIM.2013.2293 222

[5] De Vos M, Kroesen M, Emkes R, et al. P300 speller BCI with a mobile EEG system: Comparison to a traditional amplifier Journal of Neural Engineering 2014;**11**:036008. DOI: 10.1088/1741-2560/11/3/036008

[6] Gallager RG. Low-Density Parity-Check Codes. Cambridge, MA: MIT Press; 1963

[7] Limpaphayom P, Winick KA. Power- and bandwidth-efficient communications using LDPC codes. IEEE Transactions on Communications. 2004;**52**(3):350-354. DOI: 10.1109/TCOMM.2004.823565

[8] Franceschini M, Ferrari G, Raheli R. LDPC Coded Modulations. Springer; Heidelberg, 2009. pp.37-59. DOI: 10.1007/978-3-540-69457-1_3

[9] Ohtsuki T. LDPC codes in communications and broadcasting. IEICE Transactions on Communications. 2007;**E90-B**(3):440-453. DOI: 10.1093/ietcom/e90-b.3.440

[10] Boroujeny BF. OFDM versus filter bank multicarrier. IEEE Transactions on Signal Processing. 2011;**28**(3):92-112. DOI: 10.1109/MSP.2011.940267

[11] Bouhadda H, Shaiek H, Roviras D, et al. Theoretical analysis of BER performance of nonlinearly amplified FBMC/OQAM and OFDM signals. EURASIP Journal on Advances in Signal Processing. 2014;**2014**:60. DOI: 10.1186/1687-6180-2014-60

[12] Caus M, Ana I PN, Kliks A. Characterization of the effects of multi-tap filtering on FBMC/OQAM systems. EURASIP Journal on Advances in Signal Processing. 2014;**2014**:84. DOI: DOI: 10.1186/1687-6180-2014-84

[13] Bellanger M, Ruyet DL, Roviras D, et al. Fbmc physical layer: A primer. PHYDYAS. 2010

[14] Lin CF. Mobile telemedicine: A survey study. Journal of Medical Systems. 2012;**36**(2):511-520. DOI: 10.1007/s10916-010-9496-x

[15] Lin CF. An advance wireless multimedia communication application: Mobile telemedicine. WSEAS Transactions on Communications. 2012;**9**(3):206-215

[16] Lin CF, Hung SI, Chiang IH. An 802.11n WLAN transmission scheme for wireless telemedicine applications. Proceedings of the Institution of Mechanical Engineers. Part H, Journal of Engineering in Medicine. 2010;**224**(10):1201-1208. DOI: 10.1177/0954411911434246

[17] Lin CF, Li CY. A DS UWB transmission system for wireless telemedicine. WSEAS Transactions on Systems. 2008;**7**(7):578-588

[18] Lin CF, Chang WT, Lee HW, et al. Downlink power control in multi-code CDMA mobile medicine system. Medical & Biological Engineering & Computing 2006;**44**:437-444. DOI: 10.1007/s11517-009-0458-8

[19] Lin CF, Chang KT. A power assignment mechanism in Ka band OFDM-based multi-satellites mobile telemedicine. Journal of Medical and Biological Engineering. 2008;**28**(1):17-22

http://dx.doi.org/10.5772/intechopen.68838

[20] Lin CF, Chen JY, Shiu RH, et al. A Ka band WCDMA-based LEO transport architecture in mobile telemedicine. In: Martinez L, Gomez C, editors. Telemedicine in the 21st Century. Nova Science Publishers; USA, 2008. pp. 187-201

[21] Lin CF, Wang SE, Lu YC, et al. Mobile cloud-based blood pressure healthcare for education. In: Bonney W, editor. Mobile Health Technologies-Theories and Applications. Intech Science Publishers; Croatia, 2016. pp. 99-114

Chapter 6

Review of Artifact Rejection Methods for Electroencephalographic Systems

Suguru Kanoga and Yasue Mitsukura

Additional information is available at the end of the chapter

http://dx.doi.org/10.5772/68023

Abstract

Technologies using electroencephalographic (EEG) signals have been penetrated into public by the development of EEG systems. During EEG system operation, recordings ought to be obtained under no restriction of movement for routine use in the real world. However, the lack of consideration of situational behavior constraints will cause technical/biological artifacts that often mixed with EEG signals and make the signal processing difficult in all respects by ingeniously disguising themselves as EEG components. EEG systems integrating gold standard or specialized device in their processing strategies would appear as daily tools in the future if they are unperturbed to such obstructions. In this chapter, we describe algorithms for artifact rejection in multi-/single-channel. In particular, some existing single-channel artifact rejection methods that will exhibit beneficial information to improve their performance in online EEG systems were summarized by focusing on the advantages and disadvantages of algorithms.

Keywords: electroencephalographic signal, artifact rejection, blind source separation, regression, filtering, signal decomposition, non-negative matrix factorization

1. Introduction

Variegated branching patterns and trends of sympathetic neurons for realizing the brain function/dysfunction have yet to be completely definitized so far. A functional neuroimaging technique of the human brain has established itself as a trustworthy visible tool to definitize indeterminate patterns and discover new functions [1]. Indeed, visualized information through neuroimaging techniques has contributed building intuitive understanding and relative quantification of brain functions [2, 3].

Key benefits of the electroencephalographic (EEG) modality hold over other neuroimaging techniques (e.g., local field potential, near infrared spectroscopy, and electrocorticogram) are the high temporal resolution on the order of milliseconds, the small installation space for operating systems, and its usability in noninvasive recording [4]. Although the spatial resolution and specificity are low because it observes the volume conduction effects in brain network [5], this has been attracted attention as a viable and inexpensive modality to study kaleidoscopic functional states of the cerebral cortex: where, when, how, and under what our brain functions come into being [6]. Therefore, providing a capacity to adapt EEG systems to real environments is always a major challenge for neuroscientists and neuroengineers on the final stretch of constructing systems.

Using an extremely small number of electrodes (the single-electrode case would be an extreme case) for signal acquisition should result in better practical application in daily life. Recently, specialized (headband type or headset type) devices, which are endowed with small number of electrodes less than gold standard devices having 16, 32, 64, or more channels, have been developed as for compact, portable, and feasible EEG systems to use themselves in the real environments [7]. The devices are usually implemented with dry electrodes and wireless sensor network technology for recordings. These can diminish the burden on the user caused by oppressive feeling in the head, eliminate the discomfort from conductive gel or paste, and improve degree of freedom of movements by doing away with wires plugged into an amplifier [8].

However, technical/biological artifacts, such as active power line interference, eyeblink, and muscle activity caused by recording mistake, good conductivity of the scalp, and so on, are often mixed with EEG signals whether the type of device is gold standard or specialized. They ingeniously disguise themselves as EEG components in observed EEG signals and cause a discrepancy between research motivation and system realization. Removing mimetic components (artifacts) or extracting intrinsic EEG components from observed EEG signals will become a more important process in all EEG systems for practical use even if single electrode is integrated with data acquisition module by a specialized device.

Disclosing the meaning of electric signals comprising various neuronal populations (sources) breaks down the EEG inverse (blind source separation (BSS)) problem [9]. It is well known that the enormous indeterminacies in brain make the BSS problem ill-posed; however, statistical natures lead to restoring the well-posedness of the problem in a biosignal processing. By the properties, theoretically multivariate statistical analysis approaches like independent component analysis (ICA) can separate observed EEG signals into spatially and temporally distinguishable components effectively, and then, estimated components will be identified as neuronal or artifactual sources by hard/soft threshold to reconstruct artifact-free EEG matrix [10, 11]. Whereas there are several reviews on artifact rejection methods including overall procedure (signal separation, component identification, and signal reconstruction) for multi-channel EEG signals [12–16], we have never seen review of artifact rejection methods for single-channel EEG signals. In this chapter, we therefore describe algorithms for artifact rejection in multi-/single-channel EEG signals.

2. Concise description of technical/biological artifacts

2.1. Technical artifacts

Technical artifacts such as power line interference, impedance fluctuation, and wire movement superimpose their energy on observed EEG signals because of faults in setting conditions

[18, 19]. These can be precluded from easy ways, detaching a charging AC adapter from the recording device, carefully attaching electrodes to the scalp, and using appropriated electrode wires or adhesive tapes to stabilize wires shown in **Figure 1**. The cross mark in the figure indicates detaching the source of technical artifact from the setting conditions.

2.2. Biological artifacts

Biological artifacts, which are discharged potentials of internal organs, diffuse their energy over the head and reach each electrode attaching on the surface of the scalp as observed EEG signal. They contaminate observed signals due to the iron accumulation in the brain and good conductivity of the scalp can be broadly separated into four categories: (i) muscular, (ii) cardiac, (iii) eyemovement, and (iv) eyeblink. EEG devices capture comprehensive electric field which was reached at an electrode even if the potential contains information of electrophysiological actions except neuronal one (see **Figure 2**). Because all electrical potentials will be equally and blindly treated, recording information including only EEG components from electrodes placed on the scalp is hardly realized. Furthermore, frequency characteristics of biological artifacts and neuronal oscillations could be overlapped. That means that shunning contact with biological artifacts may seem hopelessly difficult compared with technical

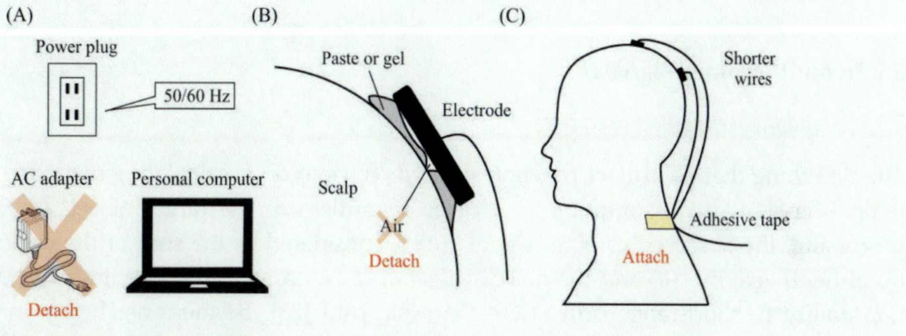

Figure 1. Ways of precluding technical artifacts [17]. (A) Power line interference. (B) Impedance fluctuation. (C) Wire movement.

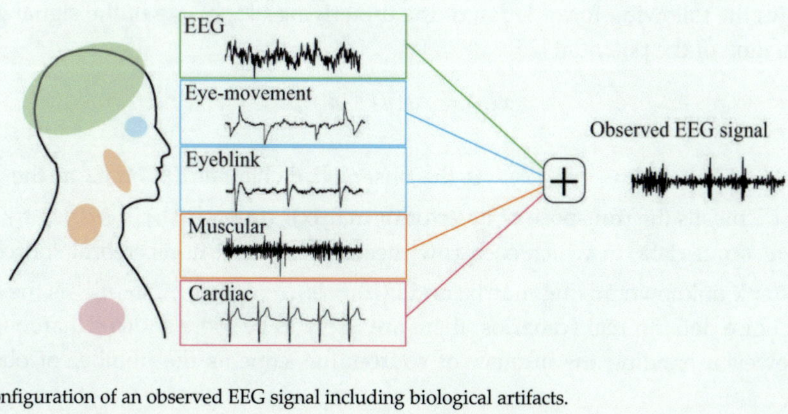

Figure 2. Configuration of an observed EEG signal including biological artifacts.

artifacts. If contaminated epochs are found in visual or quantitative analysis, the EEG system has to ignore them before deciding control commands. Otherwise, the operator will make a fatal mistake in its system by counterfeit EEG patterns [12, 17].

Alternatively, signal processing techniques can extract EEG components from observed signals. Through this process, EEG systems would provide correct outputs for their unique and beneficial interface. Even today, many works for detection, classification, and removal of artifacts within observed EEG signals have been reported [20–22].

3. Review of existing methods on artifact rejection

In this section, the standard assumptions of observed cerebral signal for spatially and temporally separating components are described before introduction of artifact rejection methods to reach deep understanding of the statistical framework. Then, methods of multi-/single-channel artifact rejection (principal component analysis (PCA), independent component analysis (ICA), regression, filtering, ICA-based signal decomposition, and nonnegative matrix factorization) are presented. Each algorithm has specialized approaches for calculating demixing matrix, identifying separated components, and denoising the artifactual components to complete source separation. We have focused on the advantages and disadvantages of approaches.

3.1. In multi-channel signals

3.1.1. Standard assumption of sources

The first thing that all artifact rejection methods have to do is calculating demixing matrix $\boldsymbol{W}$ under the standard assumption of sources regardless of the target object. In EEG signal processing, the observed cerebral signal $x(n)$ is considered as the sum of the cerebral source (local-field) activity $s(n)$ and the noise/artifact $d(n)$. Neuronal cells have limited their connection ability to short-range order (less than 500 μm) [23]. Besides, synchrony in local-field activities diffuses through a contiguous cortical area rather than jump between distant and weakly connected cortical areas [24].

Therefore, an assumption that cerebral sources and non-cerebral sources are linearly combined, allows the following formulation of the underlying biophysics of the signal generation and propagation of the potential [25]:

$$\boldsymbol{x}(n) = \boldsymbol{As}(n) + \boldsymbol{d}(n), \tag{1}$$

where: $\boldsymbol{x}(n) = [x_1(n), x_2(n), \ldots, x_P(n)]^{\mathrm{T}}$ is the observed P-channel EEG data at the n-th point (superscript T means the transpose of a vector or matrix); $\boldsymbol{s}(n) = [s_1(n), s_2(n), \ldots, s_Q(n)]^{\mathrm{T}}$ is the Q unknown source data, in which each row means cerebral or non-cerebral source; $\boldsymbol{A}$ is the $P \times Q$ full-rank unknown mixing matrix; and $\boldsymbol{d}(n) = [d_1(n), d_2(n), \ldots, d_P(n)]^{\mathrm{T}}$ is the P additive zero-mean noise data. In real scenarios, there are likely to be more sources than observations ($Q > P$); however, handing the number of sources the same as the number of observations

($Q = P$) does not normally become a fatal problem. Thus, most algorithms extract a linear combination of sources belonging to the same subspace [26, 27].

All algorithms have a common disadvantage that they can only handle over-determined mixture for the inverse process while having no priori information on the characteristics of the sources. Additional three assumptions are reluctantly accepted: (i) the noise/artifact is spatially uncorrelated with the observed data ($\mathbb{E}[\boldsymbol{A}\boldsymbol{s}(n)\boldsymbol{d}(n)^{\mathrm{T}}] = \boldsymbol{0}$, where $\mathbb{E}[\cdot]$ is the expectation operator), and temporally uncorrelated ($\mathbb{E}[\boldsymbol{d}(n)\boldsymbol{d}(n+\tau)^{\mathrm{T}}] = \boldsymbol{0}$, where τ is lag time and $\forall\tau > 0$); (ii) the number of sources is equal to or less than the number of observations ($Q \leq P$); and (iii) the mixing matrix A is stationary [28].

3.1.2. Blind source separation algorithms

Under aforementioned assumptions, BSS approaches estimate sources $\hat{\mathbf{S}} = [\hat{\boldsymbol{s}}(1), \ldots, \hat{\boldsymbol{s}}(N)]$ from observed EEG data $\boldsymbol{X} = [\boldsymbol{x}(1), \ldots, \boldsymbol{x}(N)]$. Unsupervised learning methods such as PCA and ICA jointly estimate demixing matrix $\boldsymbol{W}\ (= \boldsymbol{A}^{-1})$:

$$\hat{\boldsymbol{s}}(n) = \boldsymbol{W}\boldsymbol{x}(n). \tag{2}$$

Each unsupervised learning method has an algorithm that is subject to various indices: uncorrelatedness, independence, non-Gaussianity, instantaneous propagation, and linearity [29]. Linear mixture concept of blind EEG source separation is shown in **Figure 3** that presents a demixing matrix $\boldsymbol{W}(=\boldsymbol{W}_1\boldsymbol{W}_2)$ as two-step estimator because some methods firstly decorrelate an observed matrix by $\boldsymbol{W}_1$ and then demix it by $\boldsymbol{W}_2$. Given a mixing matrix $\boldsymbol{A}$ is composed of the three blind cerebral sources $\boldsymbol{s}(n)$ and provides the same number of observations $\boldsymbol{x}(n)$ in the figure.

PCA converts the observed matrix of possibly correlated variables into values of linearly uncorrelated variables (principal components (PCs)) with the first-and second-order statics [30]. This algorithm conducts the eigenvalue decomposition to get the directions $\boldsymbol{u}$ of greater

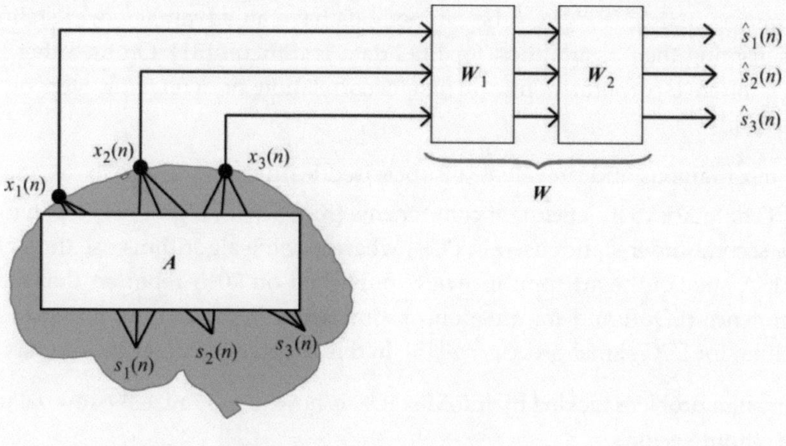

Figure 3. Linear mixture concept of blind EEG source separation [15, 17].

variance in the input space of the EEG data $\boldsymbol{X}$ based on assumptions that data are jointly normally distribution, and the sources are uncorrelated. In order to satisfy the assumptions, obtained matrix $\boldsymbol{X}^{\text{old}}$ should be standardized to decorrelate samples of the same dimension ($\mathbb{E}[\boldsymbol{x}(n)\boldsymbol{x}(n+\tau)^{\text{T}}] = \boldsymbol{0}$) and to uniform unit ($\mathbb{V}[\boldsymbol{X}_p] = 1$).

In PCA algorithm, the first PC, which has the largest variance in the standardized input space, is a linear combination of $\boldsymbol{X}$ defined by weights $\boldsymbol{u}_1 = [u_1, \ldots, u_P]^{\text{T}}$:

$$\text{PC}_1 = \boldsymbol{X}^{\text{T}}\boldsymbol{u}_1, \tag{3}$$

$$\mathbb{V}[\text{PC}_1] = \mathbb{V}[\boldsymbol{X}^{\text{T}}\boldsymbol{u}_1] = \boldsymbol{u}_1{}^{\text{T}}\sum \boldsymbol{u}_1, \tag{4}$$

where $\sum$ $(= \boldsymbol{X}\boldsymbol{X}^{\text{T}}/(N-1))$ is covariance matrix of $\boldsymbol{X}$. Therefore, this algorithm formulates the given problem in an optimization problem:

$$\max \boldsymbol{u}_1{}^{\text{T}}\sum \boldsymbol{u}_1, \tag{5}$$

$$\text{subject to } \boldsymbol{u}_1{}^{\text{T}}\boldsymbol{u}_1 = 1. \tag{6}$$

It can be solved by Lagrange multiplier method:

$$L(\boldsymbol{u}_1, \lambda_1) = \boldsymbol{u}_1{}^{\text{T}}\sum \boldsymbol{u}_1 + \lambda_1(1 - \boldsymbol{u}_1{}^{\text{T}}\boldsymbol{u}_1), \tag{7}$$

$$\frac{\partial L(\boldsymbol{u}_1, \lambda_1)}{\partial \boldsymbol{u}_1} = 2\sum \boldsymbol{u}_1 - 2\lambda_1\boldsymbol{u}_1 = 0, \tag{8}$$

$$\boldsymbol{u}_1{}^{\text{T}}\sum \boldsymbol{u}_1 = \lambda_1\boldsymbol{u}_1{}^{\text{T}}\boldsymbol{u}_1 = \lambda_1. \tag{9}$$

The covariance matrix $\sum$ is sequentially decomposed into eigenvector $\boldsymbol{u}_p$ and eigenvalue λ_p by an assumption that the PCs are orthogonal. The eigenvector $\boldsymbol{u}_p$ is similar to the column of the inverse demixing matrix $\boldsymbol{W}^{-1}$. PCA-based methods have an advantage over stationary data; however, satisfying their assumption for EEG data is difficult [31]. On the other hand, PCA algorithm is often incorporated into a first decorrelation or whitening step of some ICA algorithms [32].

ICA is the most famous and prevalent unsupervised learning algorithm to decompose multi-channel EEG data $\boldsymbol{X}$ into independent components (ICs) $\hat{\boldsymbol{S}}$ with high-order (spatial) moments, beyond the second-order statics used in PCA, whereas some algorithms use the statics as well as PCA [4]. A state-of-the-art topical review published on 2015 reported that second order blind interference (SOBI) and information maximization (InfoMax) are the most commonly used algorithm for EEG signal processing [15]. In this chapter, we describe InfoMax algorithm.

The fundamental problem tackled by InfoMax ICA is how to minimize the mutual information (MI) of the output vector $\hat{\boldsymbol{s}}$,

$$\mathrm{MI}(\hat{\mathbf{s}}) = \sum_{p=1}^{P} H(\hat{\mathbf{s}}_p) - H(\hat{\mathbf{s}}). \tag{10}$$

Probability density functions of observed signal $p(\mathbf{x})$ and estimated signal $p(\hat{\mathbf{s}})$ have following relationship:

$$p(\hat{\mathbf{s}})d\hat{\mathbf{s}} = p(\mathbf{x})d\mathbf{x}, \tag{11}$$

$$d\hat{\mathbf{s}} = J(\mathbf{x})d\mathbf{x} = |\mathbf{W}|d\mathbf{x}, \tag{12}$$

$$p(\hat{\mathbf{s}}) = p(\mathbf{x})d\mathbf{x} = p(\mathbf{W}^{-1}\hat{\mathbf{s}})|\mathbf{W}|^{-1}, \tag{13}$$

where $J(\mathbf{x})$ is Jacobian matrix. The estimating entropy $H(\hat{\mathbf{s}})$ is given by:

$$\begin{aligned} H(\hat{\mathbf{s}}) &= -\int p(\hat{\mathbf{s}}) \log p(\hat{\mathbf{s}}) d\hat{\mathbf{s}} \\ &= -\int \Big(\log p(\mathbf{x}) - \log|\mathbf{W}|\Big) p(\mathbf{x}) d\mathbf{x} \\ &= -\int p(\mathbf{x}) \log p(\mathbf{x}) d\mathbf{x} + \log|\mathbf{W}| \\ &= H(\mathbf{x}) + \log|\mathbf{W}|. \end{aligned} \tag{14}$$

Therefore, the MI can be rewritten as following:

$$\mathrm{MI}(\hat{\mathbf{s}}) = \sum_{p=1}^{P} H(\hat{\mathbf{s}}_p) - H(\mathbf{x}) - \log|\mathbf{W}|. \tag{15}$$

By partially differentiating this index on parameters $\mathbf{W}$, optimized solution for source separation will be obtained.

$$\frac{\partial \mathrm{MI}(\hat{\mathbf{s}})}{\partial \mathbf{W}} = \sum_{p=1}^{P} \frac{\partial\Big(-\int p(\hat{\mathbf{s}}) \log p(\hat{\mathbf{s}}_p) d\hat{\mathbf{s}}\Big)}{\partial \mathbf{W}} - (\mathbf{W}^{\mathrm{T}})^{-1} = -\mathbb{E}[\boldsymbol{\varphi}(\hat{\mathbf{s}})\mathbf{x}^{\mathrm{T}}] - (\mathbf{W}^{\mathrm{T}})^{-1}, \tag{16}$$

where

$$\varphi(\hat{\mathbf{s}}_p) = \frac{d \log p\,(\hat{\mathbf{s}}_p)}{d\hat{\mathbf{s}}_p}. \tag{17}$$

As analytical computation of equation as mentioned above is difficult, this algorithm uses a gradient update rule based on the natural gradient [33] and learning rate η that is a positive constant:

$$W \leftarrow W + \eta \Delta W, \tag{18}$$

$$\Delta W = \left(\mathbb{E}[\varphi(\hat{s})x^{\mathrm{T}}] + (W^{\mathrm{T}})^{-1}\right)W^{\mathrm{T}}W = \left(\mathbb{E}[\varphi(\hat{s})\hat{s}^{\mathrm{T}}] + I\right)W. \tag{19}$$

3.1.3. Component identification after source separation

After source separation, estimated sources $\hat{S}$ have to be continuously identified as neuronal or artifactual sources to reconstruct artifact-free EEG matrix $\hat{X}$. Visual inspection of scalp topography and empirical judgment was given the credit for identification of components [10, 14]. The overused techniques are still examined in an expedient manner for checking the results. That leads to increase in workload; therefore, hard/soft-threshold function, probability approach, and machine learning algorithm with features of the prepared material have been used for automatically identifying artifacts in estimated sources to reduce the workload and to get more repeatable labels [34, 35]. Proposing automatic and unsupervised component identification algorithm to characterize more precisely and flexibly has still been an active research area [36, 37]. Once estimated sources are identified, they advance to next step called denoising step, and then an underlying EEG matrix will be reconstructed using inverse linear demixing process (see **Figure 4**).

3.2. In single-channel signals

3.2.1. Discrepancy among standard assumptions about multi-/single-channel data

We can easily imagine that single-channel data do not always satisfy the assumptions for BSS techniques. Calculating demixing matrix W is especially difficult with single-channel artifact rejection methods (see **Figure 5**), so that researchers are forced to select whether to add information by using the reference channel before applying a method or to separate data by using only one-channel.

3.2.2. Regression

Regression algorithm was most frequently used to remove artifact up to the mid-1990s [38, 39]. In this algorithm, an observed EEG signal $x(n)$ can be expressed as

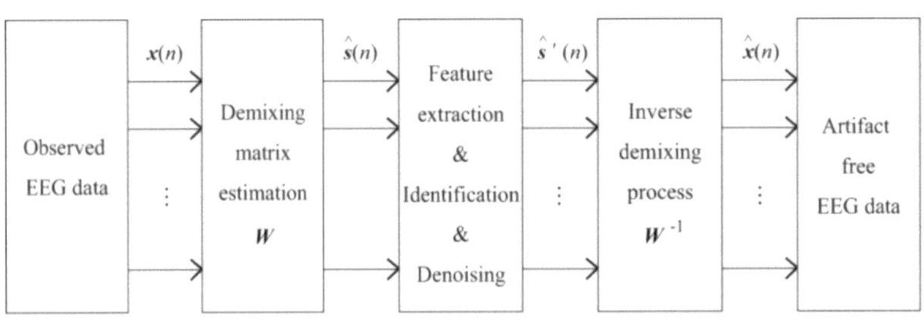

Figure 4. Block diagram of the blind source separation [11].

http://dx.doi.org/10.5772/68023

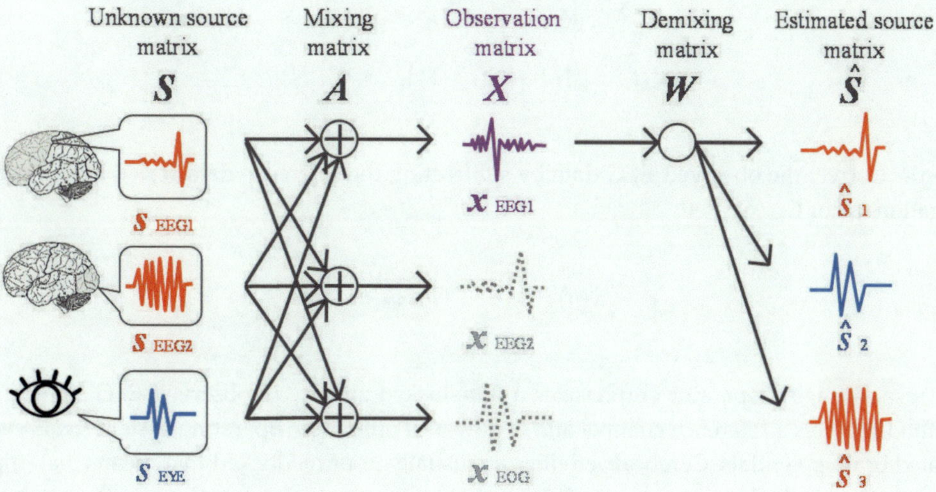

Figure 5. Procedure of signal separation in single-channel artifact rejection methods.

$$x(n) = x_{\mathrm{EEG}}(n) + x_{\mathrm{Art}}(n) + d(n), \tag{20}$$

where $x_{\mathrm{EEG}}(n)$, $x_{\mathrm{Art}}(n)$, and $d(n)$ are intrinsic EEG data, artifact, and noise. It is assumed that the expected value of $d(n)$ is 0.

The artifact would be corrected by calculating propagation factors to estimate the relationship between the reference signal $x_{\mathrm{Ref}}(n)$ and the observed EEG signal and subtracting the regressed portion [40]. The rationale of the procedure is as follows:

Step 1. Separately average over observed EEG and reference signals of T trials to estimate the artifact waveform related variation for the channels:

$$\overline{x}(n) = \frac{1}{T}\sum_{t=1}^{T} x_t(n), \tag{21}$$

Step 2. Subtract the averages from every trial data to obtain deviations:

$$\boldsymbol{x}'(n) = \boldsymbol{x}(n) - \overline{\boldsymbol{x}}(n), \tag{22}$$

where $\overline{x}(n)$ is duplicated $T \times 1$ matrix of the observed EEG average,

Step 3. Calculate the propagation factor C by linear least-square regression whereby the observed EEG data are considered as a dependent variable and the reference data are considered as the independent variable:

$$\boldsymbol{X} = C(\boldsymbol{X}_{\mathrm{Ref}}), \tag{23}$$

where

$$\mathbf{X} = [\mathbf{x}'(1), \ldots, \mathbf{x}'(t), \ldots, \mathbf{x}'(T)]^{\mathrm{T}}, \tag{24}$$

$$\mathbf{x}'(t) = [x'\left(1 + N(t-1)\right), \ \ldots, \ x'(tN)], \tag{25}$$

Step 4. Correct the observed EEG data by subtracting the reference data scaled by the propagation factor C:

$$\hat{x}(n) = x(n) - C\left(x_{\mathrm{Ref}}(n)\right). \tag{26}$$

Because averaging operator emphasizes a time-locked activity in observed EEG signals, this method requires a reference channel and is powerful only if the operating system treats event-related brain potentials. Cerebral activities are usually not time-locked that means that important nontime-locked components will be lost by the averaging operation. Furthermore, this method does not take bidirectional contamination into account and cancels the cerebral information from each observed EEG signal upon linear subtraction [41]. Despite its disadvantages, regression is still used as the "gold standard" method to which the performance of any artifact rejection algorithms may be compared.

3.2.3. Filtering

Band-pass is one of the classical and simple separation attempts to remove artifacts from an observed EEG signal. This method is effective if the spectral distributions of the EEG component and artifact do not overlap, and there are small band artifacts such as power line noise (50/60 Hz interference) [42]. However, fixed-gain filtering is not effective for biological artifacts because it will attenuate EEG component and change both amplitude and phase of signal if the filtering keeps doing that [43]. Some adaptive algorithms try to adapt the filter parameters $\boldsymbol{w}$ to minimize the error between the artifact-free EEG signal $\hat{\boldsymbol{x}}(n)$ and the desired original signal $\boldsymbol{x}(n)$ to suppress the limitations of this method.

Adaptive filtering assumes that the intrinsic EEG signal and artifact are uncorrelated; therefore, the artifact is considered to be an additive noise within the observed signal:

$$x_t(n) = s_t(n) + n_{0t}(n), \tag{27}$$

where $x_t(n)$ is the observed EEG signal of t-th trial, $n_0(0)$ is the additive noise to offset and is uncorrelated with intrinsic EEG signal $s_t(n)$. The filter parameters w are iteratively adjusted by a feedback (recursive) process designed to make the output as close as possible to some desired response with an additive noise interference [44, 45]. **Figure 6** shows the noise canceller system using adaptive filtering. In this system, the primary input $x_t(n)$ and the reference input $x_{\mathrm{Ref}t}(n)$ are the observed EEG and reference signals. A reference input $x_{\mathrm{Ref}t}(n) = n_{1t}(n)$ which is a noise correlated with $n_{0t}(n)$ and uncorrelated with intrinsic EEG signal $s_t(n)$, adds information to minimize the error $e_t(n)$ between the response $y_t(n)$ and the desired response.

http://dx.doi.org/10.5772/68023

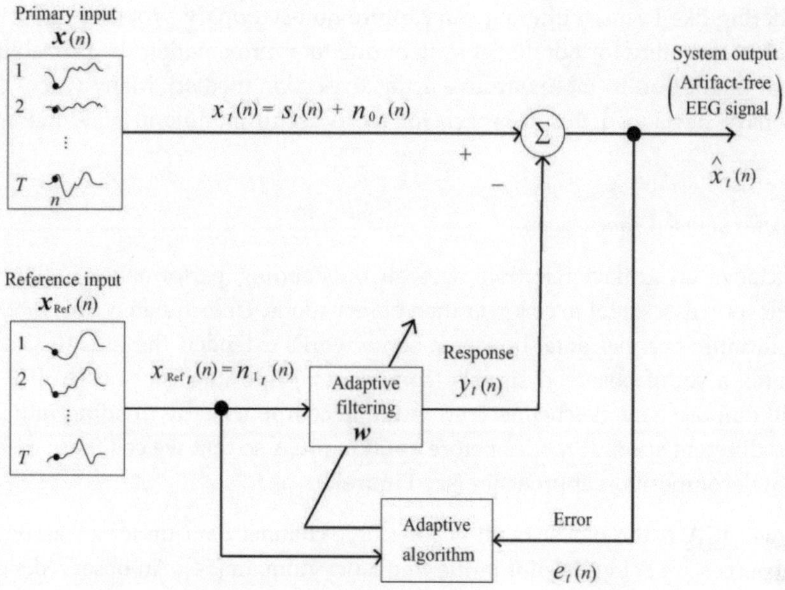

Figure 6. Noise canceller system using adaptive filtering [17, 47].

Recursive least squares (RLS)-based adaptive filtering presents a superior performance than least mean squares-based one [46]. The algorithm can be implemented using the following equations:

$$\boldsymbol{g}(n) = \frac{\boldsymbol{R}(n-1)\boldsymbol{x}_{\mathrm{Ref}}(n)}{\lambda + \boldsymbol{x}_{\mathrm{Ref}}{}^{\mathrm{T}}(n)\boldsymbol{R}(n-1)\boldsymbol{x}_{\mathrm{Ref}}(n)}, \tag{28}$$

$$\boldsymbol{e}(n) = \boldsymbol{x}(\mathrm{n}) - \boldsymbol{y}(\mathrm{n}), \tag{29}$$

$$\boldsymbol{y}(n) = \boldsymbol{w}(n)\boldsymbol{x}_{\mathrm{Ref}}(n), \tag{30}$$

$$\boldsymbol{R}(n) = \frac{\boldsymbol{R}(n-1) - \boldsymbol{g}(n)\boldsymbol{x}_{\mathrm{Ref}}{}^{\mathrm{T}}(n)\boldsymbol{R}(n-1)}{\lambda}, \tag{31}$$

$$\boldsymbol{w}(n) = \boldsymbol{w}(n-1) + \boldsymbol{g}(n)\boldsymbol{e}(n), \tag{32}$$

where $\boldsymbol{g}(n)$ and $\boldsymbol{w}(n)$ are the gain vector and the filtering parameters. The initial value of cross-correlation $\boldsymbol{R}(0)$ is $\delta\boldsymbol{I}$, where δ and $\boldsymbol{I}$ are some sufficiently large positive value and identity matrix. The updated filter parameters lead to output artifact-free EEG signal.

Consequently, adaptive filtering approach has a potential to recover "pure" EEG signal more rapidly and accurately than linear regression for ocular and cardiac artifacts [48]. However, it is rather difficult to converge to the solution of filtering parameters if muscular and vibration artifacts have contaminated in the observed EEG signal. In that situation, the algorithm sometimes does not converge because of their convulsive burst.

Optimal filtering like Kalman filtering can capture non-stationary properties of artifacts. The framework has flexibility for non-linear system due to approximating the probability density function that might lead to more effective artifact rejection method. Many works on filtering algorithms have developed this approach for more useful module in real-time applications [49, 50].

3.2.4. ICA-based signal decomposition

ICA will achieve an artifact rejection with an outstanding performance if the number of independent sources is equal to or lower than observations. Unfortunately, this method is only applicable to multi-channel data; however, some works extended the idea to single-channel data to unmix a set of observed signals (components) into intrinsic sources [51–53]. These methods decompose a single-channel into multiple components by dividing into a sequence of blocks or different spectral modes before applying ICA so that we call these methods ICA-based signal decomposition approaches (see **Figure 7**).

Single-channel ICA is the oldest method for single-channel data under an assumption that stationary sources are being disjoint in the frequency domain [54]. An observed signal $x(n)$ is split up into K short segments $\mathbf{X}$, a sequence of contiguous blocks of length L which is to be handled as a set of observations.

$$\mathbf{X} = [\boldsymbol{x}(1), \ldots, \boldsymbol{x}(k), \ldots, \boldsymbol{x}(K)]^{\mathrm{T}}, \tag{33}$$

$$\boldsymbol{x}(k) = [x\left(L(k-1)+1\right), \ldots, x(kL)]^{\mathrm{T}}, \tag{34}$$

where k is the block index. A standard ICA algorithm than performs to the matrix $\mathbf{X}$ to derive the demining matrix $\mathbf{W}$. The artifacts overlap with EEG components and EEG signal has non-periodic components; therefore, this method can be applied within limited situations. Wavelet transform (WT)-based and empirical mode decomposition (EMD)-based ICA have already been reported successful in removing artifacts for solving the similar problem than single-channel ICA [42].

WT-based ICA transforms an observed signal into components of disjoint spectra (a matrix) instead of signal (a vector) via discrete WT [55].

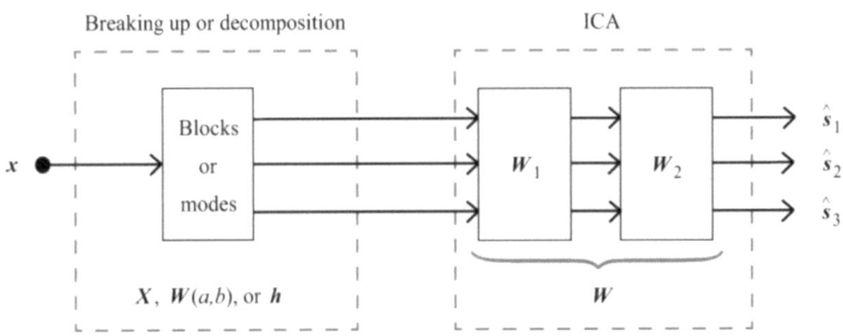

Figure 7. Procedure of separation method using ICA-based signal decomposition.

http://dx.doi.org/10.5772/68023

$$W(a,\ b) = \frac{1}{\sqrt{a}} \int x(n)\psi_{a,b}(n)dn, \tag{35}$$

$$\psi_{a,b} = \psi\left(\frac{n-b}{a}\right), \tag{36}$$

where $W(a, b)$ and $\psi_{a,\ b}$ denote that the wavelet representation of $x(n)$ and the mother wavelet with a and b defining the time-scale and location. The decision of parameters is hard if the user does not have a priori knowledge of the signal of interest. Each IC using wavelet coefficients is, respectively, identified as either neuronal or artifactual by manually. The artifactual ICs are replaced their values with arrays of zeros and then reconstructed to wavelet components. Finally, artifact-free signal is acquired by inverse discrete WT.

EMD-based ICA decomposes an observed signal into a number of K intrinsic mode functions (IMFs) $h_k(n)$,

$$x(n) = \sum_{k=1}^{K} h_k(n) + d(n), \tag{37}$$

where $d(n)$ is a residue of the original data and a nonzero mean slowly varying function with only a few or no extreme [56]. This method can remove artifacts without a priori knowledge regarding characteristics of the signal embedded in the data [57]. Each IMF has monocomponent of the original data and is estimated by an iterative process called "shifting process":

Step 1. Find the local maxima and minima in $x_k(n)$,

Step 2. Connect all of the local maxima and minima by cubic splines to form an upper and a lower envelope,

Step 3. Calculate the mean of the two envelopes, respectively,

Step 4. Obtain improved IMF $h_{k+1}(n)$ by subtracting the mean of the two envelopes from the current IMF $h_k(n)$,

Step 5. Go to Step 1 until the residue is below a stopping criterion.

This decomposition is based on the three conditions: (i) the number of extreme and the number of zero-crossing must be equal or up to plus/minus one; (ii) zero mean; and (iii) all the maxima and all the minima of IMF will be positive and negative everywhere. Each IC using IMFs is, respectively, identified as either neuronal or artifactual by manually as well as WT-based ICA. The artifactual ICs are replaced their values with arrays of zeros. Finally, reconstructed IMFs are summed simply together to acquire artifact-free signal.

WT-based and EMG-based ICA have been reported as superb methods for artifact rejection [51, 58, 59]. Therefore, a certain number of researchers tends to select them over recent years. However, separating intrinsic EEG components and artifacts are not successfully completed by this approach because frequency characteristics of biological artifacts and EEG components could be overlapped. In addition, a presence of similar oscillations in different modes or a presence of disparate amplitude oscillations in the same mode, named "mode mixing" makes the performance of artifact rejection worse [60]. Signal distortion or attenuation typically occurs according to the above-mentioned methods by excessive interference. Thus, these approaches are not suitable for real-time applications.

3.2.5. Nonnegative matrix factorization

In linear regression, filtering, and ICA-based signal decomposition approaches, parameters $\boldsymbol{W}$ cannot often converge to a solution for perfectly demixing the mixtures. This implies that partially restricting the active space should be determined for single-channel signals.

Meanwhile, non-negative matrix factorization (NMF) [61] has recently attracted attention as effective algorithms to remove artifacts from single-channel signals because it can find the latent features underlying the interactions between EEG components and artifacts. An M-dimensional non-negative data vector $\boldsymbol{x}_n$ is placed in the column of $M \times N$ matrix $\boldsymbol{X}$, where N is number of data vectors. The matrix $\boldsymbol{X}$ is based on short-time Fourier transform and approximately factorized into an $M \times K$ nonnegative matrix $\boldsymbol{H}$ and a $K \times N$ nonnegative matrix $\boldsymbol{W}$ where K is the number of "basis" which is optimized for linear approximation of the input vectors. It can be represented by the following equation:

$$\boldsymbol{x}_n \approx \boldsymbol{y}_n = \sum_{k=1}^{K} \boldsymbol{h}_k w_{k,n}, \tag{38}$$

where an $\boldsymbol{h}_k$ and a $w_{k,n}$ denote an entry of $\boldsymbol{H}$ and $\boldsymbol{W}$. This equation means that respective non-negative EEG feature (power spectrum or amplitude spectrum) vector is approximated by linear combination of the basis vector $\boldsymbol{h}_k$ weighted by the component of $w_{k,n}$. Therefore, it can be rewritten as

$$\boldsymbol{X} \approx \boldsymbol{HW}. \tag{39}$$

Some works reported that the supervised NMF could effectively factorize the observed EEG signals into the brain activity components and the artifacts if the user has artifact data in advance [62, 63]. Before applying supervised learning, template matrix $\boldsymbol{X}_{\text{Art}}$ has been factorized into $\boldsymbol{H}_{\text{Art}}$ and $\boldsymbol{W}_{\text{Art}}$. The matrix $\boldsymbol{X}$ is continuously factorized into $\boldsymbol{H}$ and $\boldsymbol{W}$ where $\boldsymbol{H}$ contains the elements of matrix $\boldsymbol{H}_{\text{Art}}$. The matrix $\boldsymbol{H}_{\text{Art}}$ has no relation to the elements of $\boldsymbol{H}$ while using standard NMF algorithm because the initial values are set randomly and updated by multiplicative rules. In supervised learning algorithm, the matrix $\boldsymbol{H}_{\text{Art}}$ is used as a fixed value that will partially restrict the active space. By contrast, activity components in the matrix $\boldsymbol{W}_{\text{Art}}$ are variable values. For this constraint, the matrix $\boldsymbol{H}$ can attempt to express EEG components in the matrix $\boldsymbol{X}$ with the remaining based K'. EEG components will be stored in the bases (see **Figure 8**).

After these processing, non-negative data of artifact-free EEG are reconstructed from the following equation:

$$\hat{X} = \boldsymbol{X} * \sum_{k=K_{\text{Art}}+1}^{K} \sum_{n=1}^{N} \frac{\boldsymbol{H}_k W_{k,n}}{HW}. \tag{40}$$

http://dx.doi.org/10.5772/68023

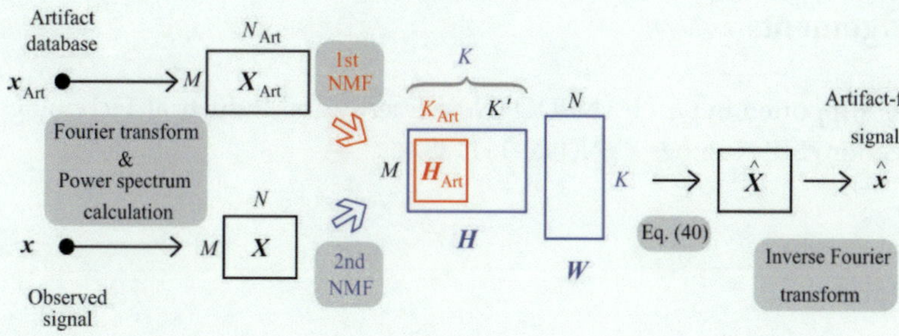

Figure 8. Procedure of supervised NMF.

Eq. (40) and inverse Fourier transform make it possible to acquire artifact-free signal. Supervised NMF is still in its infancy, showed high performance for artifact rejection. However, epoch detention step, which is not part of normal procedures in artifact rejection, must be embedded in the epoch-based method. This leads to increase the computational cost inevitably. Some low-cost (real-time) artifact detection algorithms for single-channel EEG signal [64, 65] are a silver lining in a dark cloud.

4. Conclusions

By the properties of artifacts, theoretically multivariate statistical analysis approaches such as PCA and ICA, which separate multi-channel EEG signals into spatially and temporally distinguishable components, are useful for extracting EEG components from the scalp recordings. In particular, ICA is a powerful tool for separating observed EEG signals into maximally independent activity patterns derived from cerebral or non-cerebral (artifactual) sources. However, ICA is unsuitable for analyzing EEG signals recorded by specialized EEG device because of mismatching of its assumption in the single (or few) channel case. Thus, proposing a removal method of artifact from single-channel EEG signals is currently a major challenge in EEG signal processing for the widespread use of systems as a conventional technology.

In this chapter, we tried to summarize some existing artifact rejection algorithms (PCA, ICA, regression, filtering, ICA-based signal decomposition, and NMF) focusing on the advantages and disadvantages of algorithms, which would provide beneficial information to improve their performance in online EEG systems. Last but not least, muscular artifacts reflecting body actions are natural enemies of EEG systems. The inevitable encounter must be solved by artifact rejection techniques. During real-time EEG system operation using specialized devices, unsupervised learning algorithms cannot separate observed signal into EEG and EMG components so far. Neuroscientists and neuro-engineers should carefully analyze the characteristics of artifacts and integrate them in a supervised learning algorithm for effective rejection of artifacts or extraction of intrinsic EEG components from observed EEG signals without altering the underlying brain activity to routinely use EEG systems in the future.

Acknowledgements

This work was supported in part by NEDO (New Energy and Industrial Technology Development Organization) SIP Number YYN6022-111123.

Author details

Suguru Kanoga* and Yasue Mitsukura

*Address all correspondence to: kanouga@mitsu.sd.keio.ac.jp

Keio University, Kanagawa, Japan

References

[1] Friston KJ. Modalities, modes, and models in functional neuroimaging. Science. 2009;**326** (5951):399-403. DOI: 10.1126/science.1174521

[2] Logothetis NK, Pauls J, Augath M, Trinath T, Oeltermann A. Neurophysiological investigation of the basis of the fMRI signal. Nature. 2001;**412**(6843):150-157. DOI: 10.1038/35084005

[3] Hämäläinen M, Hari R, Ilmoniemi RJ, Knuutila J, Lounasmaa OV. Magnetoencephalography—theory, instrumention, and applications to non-invasive studies of the working human brain. Reviews of Modern Physics. 1993;**65**(2):413-505. DOI: 10.1103/RevModPhys.65.413

[4] Rampil IJ. A primer for EEG signal processing in anesthesia. The Journal of the American Society of Anesthesiologists. 1998;**89**(4):980-1002

[5] Bear MF, Connors BW, Paradiso MA. Brain rhythms and sleep. In: Neuroscienece: Exploring the brain. 3rd ed. USA: Lippincott Williams and Wilkins; 1996. pp. 585-616

[6] Pfurtscheller G, Allison BZ, Brunner C, Bauernfeind G, Solis-Escalante T, Scherer R, et al. The hybrid BCI. Frontiers in Neuroscience. 2010;**4**(30):42. DOI: 10.3389/fnpro.2010.00003

[7] Liao LD, Chen CY, Wang IJ, Chen SF, Li SY, Chen BW, et al. Gaming control using a wearable and wireless EEG-based brain-computer interface device with novel dry foam-based sensors. Journal of Neuroengineering and Rehabilitation. 2012;**9**(5)DOI: 10.1186/1743-0003-9-5

[8] Chi YM, Jung TP, Cauwenberghs G. Dry-contact and noncontact biopotential electrodes: Methological review. IEEE Reviews in Biomedical Engineering. 2010;**3**:106-119. DOI: 10.1109/RBME.2010.2084078

[9] Xu P, Yao D. A novel method based on realistic head model for EEG denoising. Computer Methods and Programs in Biomedicine. 2006;**83**(2):104-110. DOI: 10.1016/j.cmpb.2006.06.002

[10] Jung TP, Makeig S, Humphries C, Lee TW, Mckeown MJ, Iragui V, et al. Removing electroencephalographic artifacts by blind source separation. Psychophysiology. 2000;**37**(2):163-178. DOI: 10.1111/1469-8986.3720163

[11] Shoker L, Sanei S, Chambers J. Artifact removal from electroencephalograms using a hybrid BSS-SVM algorithm. IEEE Signal Processing Letters. 2005;**12**(10):721-724. DOI: 10.1109/LSP.2005.855539

[12] Fatourechi M, Bashashati A, Ward RK, Birch GE. EMG and EOG artifacts in brain computer interface systems: A survey. Clinical Neurophysiology. 2007;**118**(3):480-494. DOI: 10.1016/j.clinph.2006.10.019

[13] Daly I, Nicolaou N, Nasuto SJ, Warwick K. Automated artifact removal from the electroencephalogram: A comparative study. Clinical EEG and Neuroscience. 2013;**44**(4):291-306. DOI: 10.1177/1550059413476485

[14] Romero S, Mañanas MA, Barbanoj MJ. A comparative study of automatic techniques for ocular artifact reduction in spontaneous EEG signals based on clinical target variables: A simulation case. Computers in Biology and Medicine. 2008;**38**(3):348-360. DOI: 10.1016/j.compbiomed.2007.12.001

[15] Urigüen JA, Garcia-Zapirain B. EEG artifact removal—state-of-the-art and guidelines. Journal of Neural Engineering. 2015;**12**(3):031001. DOI: 10.1088/1741-2560/12/3/031001

[16] Ghaderi F, Kim SK, Kirchner EA. Effects of eye artifact removal methods on single trial P300 detection, a comparative study. Journal of Neuroscience Methods. 2014;**221**:41-47. DOI: 10.1016/j.jneumeth.2013.08.025

[17] Kanoga S. An accurate removal of eyeblink artifact from single-channel electroencephalogram by supervised tensor factorization [dissertation]. Yokohama: Keio University; 2016. 159p. Available from: http://koara.lib.keio.ac.jp/xoonips/modules/xoonips/detail.php?koara_id=KO50002002-20164510-0003

[18] Teplan M. Fundamentals of EEG measurement. Measurement Science Review. 2002;**2**(2):1-11

[19] Huhta JC, Webster JG. 60-Hz interference in electrocardiography. IEEE Transactions on Biomedical Engineering. 1973;**20**(2):91-101. DOI: 10.1109/TBME.1973.324169

[20] Goncharova II, McFarland DJ, Vaughan TM, Wolpaw JR. EMG contamination of EEG: Spectral and topographical characteristics. Clinical Neurophysiology. 2003;**114**(9):1580-1593. DOI: 10.1016/S1388-2457(03)00093-2

[21] Olund T, Duun-Henriksen J, Kjaer TW, Sorensen HBD. Automatic detection and classification of artifacts in single-channel EEG. In: Engineering in Medicine and Biology

Society; 26-30 Aug.; Chicago, Illinois, USA. IEEE; 2014. pp. 922-925. DOI: 10.1109/EMBC.2014.6943742

[22] Hu J, Wang CS, Wu M, Du YX, He Y, She J. Removal of EOG and EMG artifacts from EEG using combination of functional link neural network and adaptive neural fuzzy inference system . Neurocomputing. 2015;**151**(1):278-287. DOI: 10.1016/j.neucom.2014.09.040

[23] Onton J, Makeig S. Information-based modeling of event-related brain dynamics. Progress in Brain Research. 2006;**159**:99-120. DOI: 10.1016/S0079-6123(06)59007-7

[24] Freeman WJ. Origin, structure, and role of background EEG activity. Part 1. Analytic amplitude. Clinical Neurophysiology. 2004;**115**(9):2077-2088. DOI: 10.1016/j.clinph.2004.02.029

[25] Sarvas J. Basic mathematical and electromagnetic concepts of the biomagnetic inverse problem. Physics in Medicine and Biology. 1987;**32**(1):11-22. DOI: 10.1088/0031-9155/32/1/004

[26] Picton TW, Bentin S, Berg P, Donchin E, Hillyard SA, Johnson R, et al. Guidelines for using human event-related potentials to study cognition: Recording standards and publication criteria. Psychophysiology. 2000;**37**(2):127-152

[27] Safieddine D, Kachenoura A, Albera L, Birot G, Karfoul A, Pasnicu A, et al. Removal of muscle artifact from EEG data: Comparison between stochastic (ICA and CCA) and deterministic (EMD and wavelet-based) approaches. EURASIP Journal on Advances in Signal Processing. 2012;**2012**:127. DOI: 10.1186/1687-6180-2012-127

[28] James CJ, Hesse CW. Independent component analysis for biomedical signals. Physiological Measurement. 2005;**26**(1):R15. DOI: 10.1088/0967-3334/26/1/R02

[29] Lee SY. Blind source separation and independent component analysis: A review. Neural Information Processing Letters and Reviews. 2005;**6**(1):1-57

[30] Berg P. Scherg M. Dipole models of eye movements and blinks. Electroencephalography and Clinical Neurophysiology. 1991;**79**(1):36-44. DOI: 10.1016/0013-4694(91)90154-V

[31] Dien J. Addressing misallocation of variance in principal components analysis of event-related potentials. Brain Topography. 1998;**11**(1):43-55. DOI: 10.1023/A:1022218503558

[32] Vigário RN. Extraction of ocular artefacts from EEG using independent component analysis. Electroencephalography and Clinical Neurophysiology. 1997;**103**(3):395-404. DOI: 10.1016/S0013-4694(97)00042-8

[33] Amari SI. Natural gradient works efficiently in learning. Neural Computation. 2006;**10**(2):251-276. DOI: 10.1162/089976698300017746

[34] Lagerlund TD, Sharbrough FW, Busacker NE. Spatial filtering of multichannel electroencephalographic recordings through principal component analysis by singular value decomposition. Journal of Clinical Neurophysiology. 1997;**14**(1):73-82

[35] Nicolaou N, Nasuto SJ. Automatic artefact removal from event-related potentials via clustering. The Journal of VLSI Signal Processing Systems for Signal, Image, and Video Technology. 2007;**48**(1):173-183. DOI: 10.1007/s11265-006-0011-z

[36] Mognon A, Jovicich J, Bruzzone L, Buiatti M. ADJUST: An automatic EEG artifact detector based on the joint use of spatial and temporal features. Psychophysiology. 2011;**48**(2): 229-240. DOI: 10.1111/j.1469-8986.2010.01061.x

[37] Mahajan R, Morshed B. Unsupervised eye blink artifact denoising of EEG data with modified multiscale sample entropy, kurtosis, and wavelet-ICA. IEEE Journal of Biomedical and Health Informatics. 2015;**19**(1):158-165. DOI: 10.1109/JBHI.2014.2333010

[38] Woestenburg JC, Verbaten MN, Slangen JL. The removal of the eye-movement artifact from the EEG by regression analysis in the frequency domain. Biological Psychology. 1983;**16**(1-2):127-147. DOI: 10.1016/0301-0511(83)90059-5

[39] Klados MA, Papadelis C, Braun C, Bamidis PD. REG-ICA: A hybrid methodology combining blind source separation and regression techniques for the rejection of ocular artifacts. Biomedical Signal Processing and Control. 2011;**6**(3):291-300. DOI: 10.1016/j.bspc.2011.02.001

[40] Gratton G, Coles MG, Donchin E. A new method for off-line removal of ocular artifact. Electroencephalography and Clinical Neurophysiology. 1983;**55**(4):468-484. DOI: 10.1016/0013-4694(83)90135-9

[41] Croft RJ, Barry RJ. EOG correction of blinks with saccade coefficients: A test and revision of the aligned-artefact average solution. Clinical Neurophysiology. 2000;**111**(3):444-451. DOI: 10.1016/S1388-2457(99)00296-5

[42] Sweeney KT, Ward TE, McLoone SF. Artifact removal in physiological signals—practices and possibilities. IEEE Transactions on Information Technology in Biomedicine. 2012;**16**(3): 488-500. DOI: 10.1109/TITB.2012.2188536

[43] Gao J, Sultan H, Hu J, Tung WW. Denoising nonlinear time series by adaptive filtering and wavelet shrinkage: A comparison. IEEE Signal Processing Letters. 2010;**17**(3):237-240. DOI: 10.1109/LSP.2009.2037773

[44] Feintuch PL. An adaptive recursive LMS filter. Proceedings of the IEEE. 1976;**64**(11):1622-1624. DOI: 10.1109/PROC.1976.10384

[45] Rupp M. A family of adaptive filter algorithms with decorrelating properties. IEEE Transactions on Signal Processing. 1998;**46**(3):771-775. DOI: 10.1109/78.661344

[46] Peng H, Hu B, Shi Q, Ratcliffe M, Zhao Q, Qi Y, et al. Removal of ocular artifacts in EEG—an improved approach combining DWT and ANC for portable applications. IEEE Journal of Biomedical and Health Informatics. 2013;**17**(3):600-607. DOI: 10.1109/JBHI.2013.2253614

[47] Rahman FA, Othman MF, Shaharuddin NA. A review on the current state of artifact removal methods for electroencephalogram signals. In: 2015 10th Asian Control Conference;

31 May–3 Jun.; Kota Kinabalu, Malaysia. IEEE; 2015. pp. 1-6. DOI: 10.1109/ASCC.2015.7244679

[48] He P, Kahle M, Wilson G, Russell C. Removal of ocular artifacts from EEG: A comparison of adaptive filtering method and regression method using simulated data. In: 27th Annual International Conference of the IEEE Engineering; 17-18 Jan.; Shanghai, China. IEEE; 2006. pp. 1110-1113. DOI: 10.1109/IEMBS.2005.1616614

[49] Morbidi F, Garulli A, Prattichizzo D, Rizzo C, Rossi S. Application of Kalman filter to remove TMS-induced artifacts from EEG recordings. IEEE Transactions on Control Systems Technology. 2008;**16**(6):1360-1366. DOI: 10.1109/TCST.2008.921814

[50] Navarro X, Porée F, Beuchée A, Carrault G. Denoising preterm EEG by signal decomposition and adaptive filtering: A comparative study. Medical Engineering and Physics. 2015;**37**(3):315-320. DOI: 10.1016/j.medengphy.2015.01.006

[51] Mijovic B, De Vos M, Gligorijevic I, Taelman J, Van Huffel S. Source separation from single-channel recordings by combining empirical-mode decomposition and independent component analysis. IEEE transactions on biomedical engineering. 2010;**57**(9):2188-2196. DOI: 10.1109/TBME.2010.2051440

[52] Lindsen JP, Bhattacharya J. Correction of blink artifacts using independent component analysis and empirical mode decomposition. Psychophysiology. 2010;**47**(5):955-960. DOI: 10.1111/j.1469-8986.2010.00995.x

[53] Khatun S, Mahajan R, Morshed BI. Comparative analysis of wavelet based approaches for reliable removal of ocular artifacts from single channel EEG. In: IEEE International Conference on Electro/Information Technology; 21-23 May; DeKalb, IL, USA. IEEE; 2015. pp. 335-340. DOI: 10.1109/EIT.2015.7293364

[54] Davies ME, James CJ. Source separation using single channel ICA. Signal Processing. 2007;**87**(8):1819-1832. DOI: 10.1016/j.sigpro.2007.01.011

[55] Lin J, Zhang A. Fault feature separation using wavelet-ICA filter. NDT & E International. 2005;**38**(6):421-427. DOI: 10.1016/j.ndteint.2004.11.005

[56] Huang NE, Shen Z, Long SR, Wu MC, Shih HH, Zheng Q, et al. The empirical mode decomposition and the Hilbert spectrum for nonlinear and non-stationary time series analysis. Proceedings of the Royal Society of London A: Mathematical, Physical and Engineering Sciences. 1998;**454**(1971):903-995. DOI: 10.1098/rspa.1998.0193

[57] Flandrin P, Goncalves P. empirical mode decompositions as data-driven wavelet-like expansions. International Journal of Wavelets, Multiresolution and Information Processing. 2004;**2**(4):477-496. DOI: 10.1142/S0219691304000561

[58] Sweeney KT, Ayaz H, Ward TE, Izzetoglu M, McLoone SF, Onaral B. A methodology for validating artifact removal techniques for physiological signals. IEEE Transactions on Information Technology in Biomedicine. 2012;**16**(5):918-926. DOI: 10.1109/TITB.2012.2207400

[59] Mourad N, Niazy RK. Automatic correction of eye blink artifact in single channel EEG recording using EMD and OMP. In: The 2013 European Signal Processing Conference; 9-13 Sept.; Marrakech, Morocco. IEEE; 2013. pp. 1-5

[60] Wu SD, Chiou JC, Goldman E. Solution for mode mixing phenomenon of the empirical mode decomposition. In: IEEE 3rd International Conference on Advanced Computer Theory and Engineering; 20-22 Aug.; Chengdu, China. IEEE; 2010. pp. 500-504. DOI: 10.1109/ICACTE.2010.5579417

[61] Lee DD, Seung HS,. Learning the parts of objects by non-negative matrix factorization. Nature. 1999;**401**:788-791. DOI: 10.1038/44565

[62] Damon C, Liutkus A, Gramfort A, Essid S. Non-negative matrix factorization for single-channel EEG artifact rejection. In: IEEE International Conference on Acoustics, Speech and Signal Processing; 26-31 May; Vancouver, BC, Canada. IEEE; 2013. pp. 1177-1181. DOI: 10.1109/ICASSP.2013.6637836

[63] Kanoga S, Mitsukura Y. Eye-blink artifact reduction using 2-step nonnegative matrix factorization for single-channel electroencephalographic signals. Journal of Signal Processing. 2014;**18**(5):251-257

[64] Majmudar CA, Mahajan R, Morshed BI. Real-time hybrid ocular artifact detection and removal for single channel EEG. In: IEEE International Conference on Electro/Information Technology; 21–23 May; DeKalb, IL, USA. IEEE; 2015. pp. 330-334. DOI: 10.1109/EIT.2015.7293363

[65] Chang WD, Cha HS, Kim K, Im CH. Detection of eye blink artifacts from single prefrontal channel electroencephalogram. Computer Methods and Programs in Biomedicine. 2016; **124**:19-30. DOI: 10.1016/j.cmpb.2015.10.011